D0948578

CONSCIENCE AND COURAGE

HOW VISIONARY CEO HENRI TERMEER BUILT A BIOTECH GIANT AND PIONEERED THE RARE DISEASE INDUSTRY

ALSO FROM COLD SPRING HARBOR LABORATORY PRESS

Abraham Lincoln's DNA and Other Adventures in Genetics

A Cure Within: Scientists Unleashing the Immune System to Kill Cancer

Is It in Your Genes? The Influence of Genes on Common Disorders and Diseases that Affect You and Your Family

Orphan: The Quest to Save Children with Rare Genetic Disorders

The Strongest Boy in the World: How Genetic Information Is Reshaping Our Lives, Updated and Expanded Edition

CONSCIENCE AND COURAGE

HOW VISIONARY CEO HENRI TERMEER BUILT A BIOTECH GIANT AND PIONEERED THE RARE DISEASE INDUSTRY

John Hawkins

CSH PRESS
COLD SPRING HARBOR LABORATORY PRESS
Cold Spring Harbor, New York • www.cshlpress.org

Conscience and Courage

How Visionary CEO Henri Termeer Built a Biotech Giant and Pioneered the Rare Disease Industry

All rights reserved
© 2019 by Odgers Berndtson, LLC
Published by Cold Spring Harbor Laboratory Press, Cold Spring Harbor, New York
Printed in the United States of America

This book has been made possible through the support of Odgers Berndtson, LLC.

Publisher	John Inglis
Director of Editorial Development	Jan Argentine
Project Manager	Inez Sialiano
Permissions Coordinator	Carol Brown
Director of Publication Services	Linda Sussman
Production Editor	Kathleen Bubbeo
Production Manager	Denise Weiss
Cover Designer	Mike Albano

Front cover: Henri Termeer photograph, with permission from Sanofi Genzyme.

Library of Congress Cataloging-in-Publication Data

Names: Hawkins, John, author.
Title: Conscience and courage : how visionary CEO Henri Termeer built a biotech giant and pioneered the rare disease industry / John Hawkins, Vice Chairman and Director, Odgers Berndtson.
Description: Cold Spring Harbor, New York : Cold Spring Harbor Laboratory Press, [2019]
Identifiers: LCCN 2019018867 (print) | LCCN 2019022245 (ebook) | ISBN 9781621823711 (ePub3) | ISBN 9781621823728 (Kindle-Mobi) | ISBN 9781621823704 (cloth)
Subjects: LCSH: Termeer, Henri A. | Genzyme (Firm) | Chief executive officers–Biography. | Biotechnology industries–Management. | Orphan drugs.
Classification: LCC HD9999.B442 (ebook) | LCC HD9999.B442 H385 2019 (print) | DDC 338.7/616151092 [B] –dc23
LC record available at https://lccn.loc.gov/2019018867
LC ebook record available at https://lccn.loc.gov/2019022245

10 9 8 7 6 5 4 3 2 1

All World Wide Web addresses are accurate to the best of our knowledge at the time of printing.

Authorization to photocopy items for internal or personal use, or the internal or personal use of specific clients, is granted by Cold Spring Harbor Laboratory Press, provided that the appropriate fee is paid directly to the Copyright Clearance Center (CCC). Write or call CCC at 222 Rosewood Drive, Danvers, MA 01923 (978-750-8400) for information about fees and regulations. Prior to photocopying items for educational classroom use, contact CCC at the above address. Additional information on CCC can be obtained at CCC Online at www.copyright.com.

For a complete catalog of all Cold Spring Harbor Laboratory Press publications, visit our website at www.cshlpress.org.

With love to my wife, Anne,
and my mother, Bette (1927–2014)

Contents

Foreword

Everyone in the room erupted into nervous laughter, all but the frowning, crossed-arms Henri Termeer in the front row, the butt of a visual joke I had just made. It was 1994 and the start of a great friendship.

The occasion was the Massachusetts Biotechnology Council annual meeting, and I was an odd choice to open it: the CEO of Vertex Pharmaceuticals, a five-year-old fledging start-up, still five years away from our first product, addressing the leadership of a rapidly growing sector led by iconic companies like Biogen, Genetics Institute (GI), and Genzyme. But GI had been bought out after a devastating patent loss and Biogen was struggling to find a breakthrough product. In contrast, Genzyme, led for 11 years by Henri Termeer, had defied scientific odds and universal skepticism to produce a medical miracle—Ceredase for the ultrarare Gaucher disease—and a bold business model for selling it. The Genzyme building across the river symbolized the business and science success to which everyone present aspired. More like a church than a factory, it had multistory, pseudo-Gothic arches framing huge, modern, steel-framed glass windows. So, I closed my talk with "the highlight of the year for Massachusetts biotech: Genzyme's new manufacturing plant!" And clicked to a colorful slide of the Taj Majal.

Everyone in the room was laughing. Except for Henri. In the years that followed, he never once mentioned this incident to me.

Before my smart-ass speech, I hardly knew him: He had status to which I could only aspire. But afterward he often reached out or made himself available to me. Sometimes it was just a short personal note, congratulating Vertex on some minor progress. Other times it was an aside at an industry gathering, a *sotto voce* comment, especially on industry-wide issues that perhaps needed collective action. Gradually, he became my primary mentor and model for how to be a public CEO.

CEOs are often advised to be circumspect about their own views when running a meeting, lest they suppress new ideas. But in any meeting with Henri in charge, it was immediately apparent what outcome he wanted, because he would state it up front. Then, quite sincerely, he would encourage dialogue and dissenting views. Of course, in the end, having been listened to, we almost always decided to do what Henri had suggested.

This respect for Henri Termeer's leadership extended beyond his fellow biotech CEOs to those in Congress involved in oversight or regulation of the drug industry or support for basic research, a cause he championed. I observed this repeatedly when I was Chair of the national Biotechnology Innovation Organization (BIO). One particularly contentious meeting in Senator Ted Kennedy's Senate inner office involved Henri, BIO staff leadership, and me. Congress was considering how to apply fairly to biologic drugs the limited but important exclusivity provided by chemical patents on small-molecule drugs. The goal across the industry, whether "small-molecule" companies like Vertex or biologics companies like Genzyme, was not advantage for one modality over the other—pills versus injections—but a predictable, level playing field for these newer biologic drugs. Senator Kennedy argued aggressively for the quick government savings that could be gained by shortening exclusivity on some of the expensive biologic blockbusters. Henri argued for approval of "follow-on biologics"—generics-like competition. Kennedy was emotional and adamant about lowering costs. Henri was steadfast and focused on the value of preserving incentives for innovation. Kennedy did not budge, expressing some anger at us, and we left the room believing we had lost the argument. But later we heard that Senator Kennedy had *immediately afterward* gone to other key Senate leaders and made the argument for innovation, repeating Henri's points. The pathway ultimately adopted (in 2010) for follow-on biologics effectively preserved innovation in the sector.

Henri Termeer was a businessman and an innovator. He led with his beliefs and his principles, none more firmly held than the obligation to keep the needs of patients first. His inspirational genius was in the ability to balance proximal concern for the patient *today* with the long-term benefits of continued innovation for all patients. He foresaw a great threat to patients if mechanisms of innovation were ineffective. That insight inspired a generation of biotech leaders, me among them, to work for both the sustainability and the systemic improvement of the innovation process when building the companies needed to alleviate the burden of human disease.

This was Henri Termeer's greatest gift. He reminded me and challenged me and mentored me in the conviction that making a drug is not enough and building a company is not enough. To these challenges he added, with energy and humility and humor, responsibility for the future of medicine. Thank you, Henri Termeer.

JOSHUA BOGER, PHD

Preface

Three primary sources gave birth to my mission of writing a book about Henri Termeer's remarkable life and leadership story. It was in first part to give back to an industry in which I have spent much of my career as an executive as well as advisor to life sciences CEOs and Board Directors; second, to express my observations about an industry chief's pioneering leadership methods and practices, which are little known and even less well-recognized; and, third, to proclaim Termeer's profound influence and the contributions he made as a founding father of biotechnology and the rare disease industry.

Henri Termeer's story is one of inspiration. It is one of the great stories of bold leadership, but also one of giving, of compassion, and of empathy. It will inform the decisions of both young and old alike, give hope to patients and their families, guide leaders, and reveal the humility of a generous, courageous man who did so much good for so many.

Perhaps the seminal moment for the production of this book was the two hours Henri Termeer and I spent together one afternoon in the winter of 2016, five months before he died unexpectedly in his Marblehead kitchen, stricken by a heart attack. At his pied-à-terre overlooking the Boston harbor, we sat in his living room, having coffee and musing about our lives and the things we found important. Termeer had invited me over just to get to know each other better.

At that meeting, I learned something I will never forget about Henri Termeer. As our meeting was concluding and I was putting on my green loden overcoat heading out the door into the blustery, cold New England winter, he mentioned, in passing, his active mentorship of 46 CEOs. He did not boast of it. But it was clearly something in which he took pride and, indeed, rightfully so. Who had ever heard of someone doing that!? I had met literally thousands of CEOs over the course of my career, many of whom had never uttered the word "mentor." It made quite an impression on me.

That was my "aha" moment, and my interest in Termeer accelerated over the following months as my curiosity grew. I learned more about his generosity and myriad contributions in realms as diverse as medicine, education, economics, and ballet. One of my early research conclusions was obvious. He was today's version of a renaissance man.

Eclipsing all others, however, as I reflected on the views I gathered through interviewing 130 people who knew or worked with him closely, I concluded that his legacy of commitment and service to patients stands out the most in its exceptionalism. Henri cared deeply about patients. Throughout his career, their well-being drove much of everything he did. As a busy, powerful corporate CEO in leading Genzyme to a listing among the Fortune 500, their well-being was, remarkably, his greatest obsession.

My mission in composing this book has been to reveal—not only through my impressions but through the words of those I interviewed—the richness of Henri Termeer's leadership principles and methods, his high character, the way he conducted himself, and how he led his life.

Termeer's leadership story is useful today and into the future. It refreshes the soul. Spanning the highs and lows of his failures and successes, Henri's story is riveting, filled with intrigue, complexity, and drama.

It is a story that needed to be told to the world, and now it has been.

J.T.W. Hawkins
Charlottesville, Virginia
April 8, 2019

CONSCIENCE AND COURAGE

HOW VISIONARY CEO HENRI TERMEER BUILT A BIOTECH GIANT AND PIONEERED THE RARE DISEASE INDUSTRY

Fortunate breaks occur when you create an environment that has ample opportunity and the foresight to capture them.

—Henri Termeer

Prologue: Patient One

Henri Termeer sat behind the wheel of his car outside Boston's renowned Brigham and Women's Hospital, waiting to pick up a batch of placentas discarded from the delivery room. At Genzyme, the two-year-old start-up he had recently joined as President, they called his little Toyota hatchback the "placentamobile."

Genzyme was working furiously on a treatment for Gaucher disease, a debilitating, often deadly, disorder involving the buildup of fatty substances in the spleen or elsewhere in the body.

Gaucher was an "orphan" disease, a rare condition that was genetic in origin and affected a small number of people who had, medically speaking, no place to call home. A majority of the orphan disease patients were children. There were no therapies available for their disorders. There were also no patient support groups or social connectivity.

But worst of all, these forgotten souls and their families were desperate because few knew much about their condition and there was insufficient incentive to cure it. The affected patient populations were too small and R&D too costly and uncertain. There was little or no hope for a cure.

Genzyme was laboriously extracting the tiny amount of enzyme found in the placentas that could break down those fatty substances, but the number of placentas needed to produce enough enzyme was beyond daunting —around 22,000 human placentas per patient per year.

On the drive back to Genzyme's headquarters, Termeer thought of the patient who was being kept alive by his precious cargo. Four-year-old Brian Berman was responding well to treatment with a reconfigured enzyme the company would later call Ceredase. Administered through painfully long intravenous infusions, it was being tested in Berman's experimental trial at the National Institutes of Health (NIH) hospital outside Washington. Biochemist Roscoe Brady, MD, was his attending physician. To keep up production of the experimental drug, the placentamobile had to stay on the road.

Henri Termeer got out of his car and carefully removed the leakproof cooler. Genzyme had leased the 15th floor of an old building in the heart of

Boston's red-light district, known as the "Combat Zone." The sight of the 37-year-old Dutchman carrying a cooler packed with medical waste did not deter the local entrepreneurs—he was propositioned three times between his car and the office.

Discovering that the building's elevator was once again broken, Henri lugged the heavy cooler up the 15 flights of stairs to the lab, where the placentas were centrifuged. As the mighty machines rumbled, forcing the liquid from the tissue that contained the enzyme, work in the entire building would stop. The floors were vibrating too much.

Meanwhile, at NIH, a different form of disruption was taking place. Brian Berman was, for the first time in his life, playing like any other "normal" little boy. A few weeks earlier, Brian had been diagnosed with Gaucher disease, a diagnosis that carried a sentence of organ failure, loss of ambulation, and likely early death. Doctors had wanted to remove his spleen. Engorged by lipids, it had enlarged so dramatically that his belly had swelled to the size of a basketball. "It was pushing all his other organs out of the way," Termeer would later recall.

Brian's mother, known professionally as Dr. Robin A. Ely, was a family physician who had given up her practice after Brian was first diagnosed with Gaucher. Small, brown-haired, and a bundle of barely suppressed energy, she discovered that the most promising work anywhere on Gaucher was taking place just down the road from her home in Potomac, Maryland.

But first, Dr. Ely and her family were told there was no hope for Brian.

"We were told he has a disease called Gaucher disease," Robin recalled. "There's no treatment and you'll have to bury your child very soon. It was awful."

Dr. Ely remembers a period of "grieving and freaking out." But then, her mother called to say she had spoken with the head of the Weizmann Institute of Science, an internationally famous research university in Israel.

"She called him and said, 'My grandchild was just diagnosed with Gaucher disease, can you help me?'" Robin remembers. "The guy said to her, 'You know what, you're very fortunate because the world's expert on Gaucher is ten minutes from where you live, and that man is Dr. Roscoe Brady.'"

"I called up his office and spoke to Dr. Brady on the phone, and I told him that I had a three-year-old son who was just diagnosed and given a death prognosis. I said, 'Do you have anything?' He said to me, 'If you knew what I had, you'd come running at my door.'

"That's what he said to me. I said, 'Well, then, here I am running at your door.' I went for a meeting with him that week.

"I said, here's what I want to do. I want to give up my practice and I want to work for you for nothing. Don't pay me a dime."

Looking back, Robin says she was acting like what is now called a "warrior mother." There was nothing she would not try. No task she would shrink from doing.

"They said, 'Okay, fine. If you think you can handle it, fine.' I went to work for them, and meanwhile Brian was getting sicker and sicker by the day. He was just about to go into heart failure, and we were going to have to do an emergency splenectomy."

She lobbied hard for her son to be included in the experimental enzyme-replacement protocol Brady was planning, pleading, "Give my child one last chance before we do this operation and remove his spleen."

It was December 15, 1983. Roscoe Brady had told Robin, "We have this modified enzyme, and your son will be the first to try it." Brian Berman was in a hospital bed and his "warrior mother" was by his side.

"We had a crash cart in the room. They didn't know whether he was going to go anaphylactic. They didn't have any idea what he was going to do."

Robin recalled unforgettably, "They injected him with the enzyme—and everything went fine. We left, and that week he started perking up. It was quite amazing."

The next week, Robin went in for a follow-up appointment and was told the doctors planned on giving Brian another injection in a month. A month? That seemed to Robin like an awfully long time.

"I don't know where I got the chutzpah," Robin recalls now, "but I said, 'you know what, I have a very strong feeling that unless you give him one injection per week you're not going to see what you're looking for. You must give it to him once a week.'

"They are looking at me like, 'who are you?' But I said, 'I am telling you, I am telling you this with all my heart. I am trying to tell you something.'"

Much to Dr. Ely's amazement, "they listened to me."

"In seven weeks his hemoglobin shot up, and his belly—it was like a balloon that somebody let the air out of. It just went ... like that. Everybody was like, 'what?' It was amazing."

But after seven weeks, for the first but not last time, Genzyme ran out of enzyme. A massive amount of placentas was needed to keep the supply on line, more than even the most well-traveled placentamobile could provide.

"They ran out," Dr. Ely says, "and they said 'we are going to try and make as much more as we can, as fast as we can.' But it was another seven weeks, and in those seven weeks he went all the way back down. It was horrible for us. It was a nightmare."

Termeer recognized the little boy's response, "When we ran out of enzyme, he would get worse pretty quick. He had a remarkably fast reaction to the enzyme."

His mother saw it too, "When they got more, they injected him. He went right back up again. Just like he had before over those first seven weeks."

To Henri Termeer, Brian Berman was "his own control." He was also all the proof he needed. In search of a cure, Termeer thought to himself, "Wow, we're there."

Brian became Patient One. He was the first to benefit from a bold, innovative new treatment for Gaucher disease. But Berman was also the first beneficiary in what would, over the next three decades, become a new paradigm that has revolutionized the biotechnology and pharmaceutical industries, and indeed the world of medicine—a paradigm built around and for the treatment of the world's rare disease patients.

Genzyme and its leader, Henri Termeer, would lead this transformation. "Our company became very purpose-driven. Other companies may be very strategy-driven... . We had a purpose," said Termeer. "The purpose was the patient. Patients were what we talked about. Patients were the pictures that we showed to each other. Patients were how we reported success of what we were doing. This connection, this thinking about the patient as being the central focus... . It is remarkable how easy that translates, how cross-cultural that becomes."

"In the rare disease world it is almost possible to know the name of every patient you are treating," Genzyme's chief medical officer Richard Moscicki, MD, once told *BioCentury* magazine.

"It becomes very personal. Patients would visit Genzyme. People had pictures of patients on their desks and in the hallways, so you knew that what you were doing was directly impacting patients you had met. That created a very different sense of mission."

Today, there are roughly 7,000 rare diseases afflicting an estimated 30 million people in the United States—or one in ten Americans. A rare disease is defined as a serious, chronic condition, often life-threatening, that affects fewer than 200,000 people. A few of the more common are well known, such as cystic fibrosis and muscular dystrophy, but the names of most are known only to those who are afflicted by them.

Add the number of patient family members, caregivers, and significant others to this population and the rare disease community expands to at least 100 million people in the United States—and well more than 500 million worldwide. The depth of feeling and urgency surrounding rare disease is anything but rare.

Today, families living with rare disease become fierce patient advocates, boosters for scientific research, activists for increased funding, and allies for other families living through similar experiences. They quit their jobs and start foundations to help fund research. They move wherever necessary to participate in clinical trials. They share their story with friends, family members, and people they work and worship with. Dealing with a rare disease becomes an all-consuming, passionate pursuit that surpasses all others.

None of this would be happening if patient families felt there was no hope. And if hope in the rare disease community has a father, it is Henri Termeer.

Henri Termeer was among the first and most successful entrepreneurs of biotechnology. He was a member of a group of gifted leaders who led fledgling, disparate businesses built on recombinant DNA technology. He pioneered the development of therapies for ultrarare diseases that not only harnessed the newest genetic technologies but were fundamentally patient-centered. Termeer was the first of biotech's leaders to be patient-centric, long before the term was "cool." He helped forge biotech's public policy agenda and inspired a generation of like-minded entrepreneurs.

Termeer was not just present at the creation of the orphan drug revolution—he was in many ways its catalyst and instigator. He took the first steps on a journey that would lead to the approval of dozens of orphan drugs and the growth of a multibillion-dollar industry and would take Genzyme from a company with 17 full-time U.S. employees to a powerhouse with more than 14,000 employees in 50 offices and labs around the world.

At the beginning of this adventure, Henri Termeer might well have felt it was just him and Brian Berman taking an enormous chance. There is always a risk in going first in a clinical trial. There is no conventional wisdom, no standard operating procedure, and no history of trial and error. But for Brian Berman and his family, it was the ultimate leap of faith.

"I can tell you that of all the things that I remember from the 30 years that I was there," Henri Termeer told an audience of business students, "those are the moments that I remember the most, and those motivated me forever the most—that moment of saying, 'Wow. It works.'

"I had so many detractors in those days, people who said you are out of your mind," he said. "But I had seen this boy."

Henri Termeer's and Genzyme's success would later be measured by the growth of Genzyme into a multibillion-dollar, Fortune 500 company and the development of a roster of innovative, life-saving treatments.

But to families of rare disease patients like Brian Berman, it all added up to one word—hope.

And, exclaiming in front of a gathering that had assembled to honor Henri Termeer, the grief-stricken father of a rare diseased daughter remembered his words the day he had learned of a therapy being developed by a Boston biotech company, "Hope was spelled Genzyme."

CHAPTER ONE

The Leader Within

Henricus Adrianus Maria Termeer was born at home to Jacques and Mary Termeer on February 28, 1946 in Tilburg, a place he described as "the eighth largest town in a very small country." Located in the south of the Netherlands, near the border with Belgium, his Dutch family could trace Termeers back 500 years. His name would later be shortened to Henri (and pronounced "**Hén**-ree" in the Dutch–English tradition) after his maternal grandfather.

Henri was the fourth of six children and the first of the Termeer siblings to be born after World War II. His parents called Henri their "liberation child."

The Netherlands had been particularly hard-hit during the war. As the conflict had intensified, 400,000 Dutch persons fled their native country seeking refuge in foreign lands. When they returned home, these refugees found a homeland left with few roads, no bridges, no railroads, and little food. The Dutch people were deeply wounded. Compounding their problems, the fleeing Germans had blown up dykes in the West, flooding parts of the countryside. Contaminated water and the specter of typhus hung over the public's consciousness.

Jacques Termeer was an opponent of the Nazi regime who took up arms against the German invaders after the Netherlands entered World War II on May 10, 1940. Jacques and Mary had been married just six weeks prior.

The German forces would quickly overrun the Dutch, routing the resistance in five days. The Dutch Commander-in-Chief, General Henri Winkelman, signed their surrender to the German invaders on May 15. After five days of conflict, 10,000 Dutch troops had been killed, injured, or declared missing.

Jacques had been captured near Eindhoven, carted off by train to Germany, and incarcerated in a Nazi prisoner-of-war camp. Living conditions in these camps were brutal. On arrival, prisoners like Jacques would have their heads shaved. They slept in hammocks in crowded barracks. Beatings with fists, clubs, and whips were not uncommon. Neither were exhaustion and sleep deprivation.

Harvesting domestic animals for their nutrition, the prisoners were referred to as "knackers." Hunger was common.

While in the POW camp, Jacques refused to work for the Germans and was made to wear the dingy blue and gray striped uniforms that became associated with that tragic chapter in history.

On the day of his capture, Jacques Termeer's family did not know if he would ever be seen again. But Jacques proved to be one of the lucky ones. Six weeks after he was captured, in observance of the Geneva Convention, the Nazis would release Jacques Termeer and the other Dutch soldiers who had been taken prisoner.

Termeer returned from Germany to Tilburg by train. He would arrive very thin, having lost one-quarter of his body weight. Mary would hardly recognize him. The first thing he did on returning home from the camp was to inspect the shop window display that had been packed up and stored.

The sign outside Heuvelstraat 39 welcomed all to The Termeer Shoe Co., the small family shoe manufacturer founded by Henri Termeer's grandfather in 1909. It was nestled between the Gimbrére clothing store and the Dreesmann department store on a bustling high street often filled with shoppers, bikes, and delivery vehicles. Termeer's store occupied the ground floor of the modest, three-story red brick building, and the family's living quarters occupied the two floors above.

During the war years, the Termeer family grew as Jacques and Mary brought two daughters and one son into this world—Ineke, Marlies, and Bert Termeer—born in 1941, 1942, and 1944, respectively. Three more boys—Henri, Paul, and Roel—would be born in 1946, 1951, and 1953.

Mary van Gorp, Henri's mother, had been born to Dutch immigrant parents in Strathmore, Alberta, Canada, a small town on the prairie outside Calgary. Henri described his maternal grandmother, Maria, as "a great adventurer."

"She did not want to sit still," he would say, "and it was this urge that convinced her husband, my grandfather whom I never met, to go on this immigration adventure and start a new life. They were very happy. My mother still recalls meeting Indians."

The family had a ranch and lived in Strathmore until Maria began to develop serious problems with her eyes when Mary was around six or seven.

"The doctors felt that the very sharp air in the Rocky Mountains was not good for her. She needed the Dutch clouds." The van Gorp family moved back to the Netherlands in 1921. Henri van Gorp had left his family's paper business when he moved to Canada, and on his return he was told, "You've had your share, and now you have to be on your own."

The elder Henri had to start again from scratch, and life was difficult. He developed health problems and died young. Mary grew up in a family that struggled with the purse. She learned from an early age how to make the most out of what she had.

In the Dutch tradition, Jacques Termeer was a mercantilist and a craftsman, a man who was absorbed in his business. He and a small staff were trained in making custom shoes; they also acquired shoes through wholesalers, selling them to their loyal customers.

The manner in which Termeer ran his shop reflected his Dutch commercial upbringing. His principles were fundamental. Your reputation was paramount. You were to avoid indebtedness, operate modestly with pureness, and conduct yourself as an individualist. In the spirit of how he ran his shop, Jacques Termeer's message to his kids and grandkids would be, "go forth, work hard, redouble your efforts."

Notwithstanding his arrival in a remarkable period of history, Henri Termeer's childhood was a traditional one. He was raised by two loving parents in a middle class, Catholic family, sustained by a father who worked diligently to provide the needed financial support and a mother who devoted her love and her life to the well-being of her children.

Later in life, Termeer would describe his mother and father in this way, "My parents were an enormous influence on all of us, my brothers and sisters and me. They were talented in dealing with kids, in giving them disciplined input while being warm at the same time. We were very fortunate."

In 1953, Jacques, Mary, and their six children would move to a new house at Burgemeester Suijsstraat 10. It was situated in a leafy residential neighborhood on the outskirts of town near Tilburg University. Henri, then seven years of age, joined the Cub Scouts. He also learned to play field hockey, a popular sport for Dutch boys as well as girls. He also occasionally played cards with his friends and family. One of his favorite games, Toepen, was a four-card Dutch version of poker.

Mary was a proud, principled Catholic mother who would sit in the front row of church every Sunday morning. Her children went to Catholic school and were expected to regularly attend Sunday services at St. Jozef Kerk, a classic nineteenth century neo-Gothic landmark, with monthly confessional also expected. Jacques and Mary would raise each of their children to become "someone," someone of substance that is. Henri's sister, Marlies, explained, "My mother's ambition was to make a great success of her children."

Henri Termeer would live his adult life with little connection to organized religion, but he no doubt drew many of his philosophies and moral foundation from these youthful impressions. His style of servant leadership, his modesty, his empathy for the plight of those less fortunate—these core traits derived, at least in part, from his faith and, by extension, his mother.

One pathway to success, for Mary Termeer, was an appreciation for art and culture—a value that would stay with Henri for the rest of his life. She played the piano, and she expected all her children to learn an instrument as well.

"Making music was always great fun for the whole group," Henri said. "I played the trumpet for a little while, not very well. I tried the violin because my father had played it. That was too hard on everyone's ears. I never became a music person. But my parents really liked the performing arts. We regularly went to concerts, ballets, and the theater as a family."

Mary Termeer impressed her children with the importance of financial self-sufficiency. Henri Termeer embraced financial independence. It yielded strength, something that would define his strategic financial approach to his first two decades at Genzyme and, in some measures, explain the way he lived his life.

After settling into their new house, his mother enrolled young Henri Termeer in a quintet of Catholic schools: first, the Montessori School for early development, which was followed by St. Thomas and St. Christopher schools for his primary education. Later, he would attend high school at Paulus Lyceum and St. Odulphus Lyceum, two of the preeminent Catholic secondary schools in Tilburg. To graduate from high school in the Dutch Catholic educational system, five years of successful study were required.

He was an average student, and yet he remained deeply connected to his education, saving every page of every report card he ever received.

As described by his siblings, Henri Termeer was an ambitious yet unexceptional young boy. He was compliant, well-mannered, smart, and ingenious, but hardly a boy wonder. He did have a deeply competitive drive that sometimes got the better of him, as his brother, Bert, recalls.

"He was one class below me in primary school, and one day there was a contest to make drawings of safety signs for people who drive cars and bicycles. I made a drawing and Henri looked at it very carefully, because sometimes he had no inspiration himself, and I had too much inspiration. So, I made a drawing of a crossroads and a sign, and he looked at it very carefully and made exactly the same drawing himself. I won first prize, and he got nothing. He was so angry! He said, 'They're the same, there's no difference, why didn't I also get first prize?'

"And then, we slept in the same room, and in the middle of the night he was still angry!"

Although Henri could be overtaken by emotion, he never stopped analyzing situations or anticipating and mitigating risk. His brother Bert remembers exploring the beach with Henri while on a family vacation. It was only a few years after the war and the boys came upon abandoned German bunkers that were just waiting to be explored.

"We were young guys and of course we were very interested to go in. I remember we climbed down there, opened the door, and we get in and it's dark, of course. We found some German newspapers, but Henri was always cautious. Not scared, but cautious. We didn't know what else we might find. Maybe the ladder would collapse, maybe there was something dangerous. He did not take the risk fully, he counted his chances before.

"That's why he always won playing cards with us," Bert Termeer says today. "He wasn't a gambler and the card game we played, Toepen, was full of gambling. He always left his money on the table so everyone could see how much he had won. In the end, he was trading like a bank. We'd say, 'Henri, can we get a loan?'"

These would be qualities that Henri would cultivate throughout his life, and not always with the intended results.

Bert continued, "He would calculate his position in every situation. So, if there's a situation, before he acts he calculates his risk. He would be quiet from a distance and then he would calculate what is his risk in this situation. His decisions are not quick, you always have to wait before Henri makes a decision."

The Termeer children were close in age and close as siblings. Once Henri borrowed his brother Bert's little Fiat 600 to get to an event at St. Odulphus and had an accident that wrecked the car. He did not have the money to pay for its repair, so Bert organized a collection in the family. Henri would later pay them back—with interest.

An early outlet for Henri's fierce determination and competitive nature was chess. He would find himself, from the age of 12, absorbed in this new pursuit to the point that it would overtake all others. He played fanatically for three years, competing in various tournaments, purchasing books, and idolizing the Dutch Grandmaster of the era, Max Euwe, the first and only Dutch World Chess Champion.

Termeer would explain his passion later, "This was more important than studying. I bought books about chess and spent a lot of time learning about the game, reading about openings, closings, and all of the different plays. When you're young, you're very impressionable. It made an impression on me.

"It got so bad that I would play chess in my head and didn't really need the board. I would sit quietly in class, so the teachers had no idea. They thought I was paying attention, but I didn't hear a word they said."

Next door to his school was a chess hall that he would frequent daily. His younger brother, Paul, recounted his prowess in this way, "His talent was to think two steps ahead, meaning strategy of course. So, he discovers he was doing this quite well, and that other people did not. He could win from people who looked more intelligent than him. And older."

In fact, Henri would only play a match he was convinced he could win. Paul remembers that "Henri knew the tricks to win. He could make others look stupid. Was he that good? He was not really that good. But he knew how to win." It was confidence building, and as became clearer with age, Henri never lacked for self-confidence.

By the time he was 15, however, it all caught up with him. He had spent the last three years absorbed in chess, tuning out lectures, avoiding his homework, and playing tournaments. Finally, his devotion to chess over his studies reached a breaking point.

Seeking to advance from the third to the fourth year of study, he twice failed examinations at Paulus Lyceum. The consequence for twice failing any grade was expulsion, but Henri's mother was not going to stand for that. She went to St. Odulphus and convinced the school principal to admit her son so that he might continue his studies there. "I know he has the skills, the proficiency," she said.

The upshot was that he would be admitted, as a result of his mother's plea, to St. Odulphus. But he would need to repeat the third year and successfully pass the examination to continue his advancement to a degree.

Then, as Henri remembered, "One day, my mother said, 'Enough is enough.' She is a magnificent woman, very decisive. I woke up one day, and my chess pieces were gone, all the books on the shelf were gone. She said, 'It's over now. No more chess.'"

She even brought a therapist to the house in case Henri had some sort of a breakdown in reaction. "They both had a temper," Ineke Termeer recalled.

Termeer never played chess again, although in his adult years, he could often be seen doodling over the chess columns in the Sunday papers, fantasizing over his long-lost pastime.

"I can tell you, Henri, to the very end of his life, every day he played chess from the paper, in his head," another family member would recall. "I would watch him every morning, with the paper open to the chess column, mentally playing it. He would not write it down, but he did it every single day."

On graduating from St. Odulphus in 1964, Termeer chose to serve what was then a mandatory obligation of military service by joining the Royal Netherlands Air Force. Termeer did his basic training and boot camp nearby in Breda, a little town 10 miles west of Tilburg, and then he was selected to attend Officer's School, which came with a longer service requirement.

"If you were a regular soldier, you had to go for 12 months, or if you became an 'under officer,' a sergeant, you stayed for 18 months," Henri would later explain. "If you became an officer, you had to stay for two years, but you got real training and were put in charge of a number of different things."

For much of his service, Henri was stationed on an airbase in The Hague. He was not chosen to fly planes but rather to make sure they not only flew well but flew on time. He was going to be a logistician, given the responsibility to oversee materials inventories and the air fleet's warehouses.

"When you're young, being asked to take responsibility for managing people or operations is a magnificent experience," Henri said. "I learned a lot. Thousands of things needed to be managed and controlled in order to keep the planes in the air.

"We had a large group of people that did this work, many of them professionals. I was in charge, and I was a kid, 19 or 20 years old. Fifty-year-old sergeants had to acknowledge me as an officer. They resisted a little bit, and tried to challenge me, to see whether I was really worthy of being the boss. It helped me build a high level of confidence at an early age. Even though I was young, I could manage and get along with people. I was able to be friendly with subordinates while maintaining good discipline.

"I had two great years in the military," he said. But this experience was bigger than implementing logistics management and inventory controls. "They were formative to me in terms of my self-confidence and I realized then that I wanted to run something. I wanted to be in charge."

Having achieved the rank of 2nd lieutenant, Termeer left the Royal Netherlands Air Force in 1966 and matriculated at Erasmus University in Rotterdam to study economics. In the Dutch educational system at that time, there was no such thing as a bachelor's degree track. Erasmus only offered an accelerated course of study leading to a master's degree, including the preparation and defense of a thesis. He lived at home, worked days in his family's shoe business, and studied in the evenings.

"Somehow," Henri would later recall, "I ended up learning something about computers. And then I went to England to write my thesis and I wrote it on a computer. This was the late '60s, and computers were big things in those early days. They couldn't fit in this room."

Henri had been placed by his school with a U.K. shoe retailer named Norvic, which had hundreds of stores around the world in Commonwealth countries—places like Canada, Australia, and South Africa. His thesis was focused on the early computerization of retail stores and the economic effects it brought, a topic he thought would help him return to the family business.

The company was so impressed with him—and his nascent computer skills—that they asked him to stay on rather than go back to school.

"My experience in the Netherlands had given me some insight into automation, and my background in economics gave me some knowledge of systems input. I wasn't a programmer, I was really a systems engineer and a manager."

He had learned systems engineering in his economics studies and in the military as well.

"Shoe manufacturing was similar," Henri recalled. "There were inventories that had to be replenished, and 25 steps in the factory production of shoes, in which all kinds of different materials are used. There were no spreadsheets then, but we invented a system for ordering and keeping an inventory of materials."

Next, his manager convinced him to implement what his plan had proposed—a massive overhaul and innovation in Norvic's systems and logistics. It was the first step the company would take to computerize itself. Termeer was in charge.

"Norvic was a relatively big firm. They had hundreds of stores. They invited three companies to help them computerize their operations because they couldn't keep track of their entire inventory.

"They put teams from each company in corners of a big conference room. I had the fourth corner. We were competing, but it was a very friendly situation. Over a period of a few weeks, each team had to develop a proposal. I gleaned what I considered the best of the best and wrote a proposal. When I handed it to management, they liked it best, and asked me to implement it.

"This was the very first time I got paid even a little bit," Henri said. "I thought it was a very interesting thing to get paid."

Henri's entrepreneurism took root during this time, as he supplemented what he was getting paid by Norvic by working as an Amway salesman. His brother, Bert, remembers visiting Henri in the United Kingdom and seeing mail deliveries full of Amway products. It was a way for him to make some extra money.

He also started to buy stocks in companies like Pan Am and a few others. His brother, Roel, who was then fulfilling his military service obligation, remembers asking his older brother for stock tips.

"After a few months in the military you get some money, and I asked Henri which stocks I should buy because I knew he was playing the market. We had learned from our father that when you have money you can make more with it, and you can do that with stocks. So, I learned from Henri how to do that when I was 18, 19 years old."

Termeer stayed with the Norvic organization for two and a half years. He moved near their manufacturing plants in East Anglia and computerized the factories. His tightly wound unit would become something similar to his own independent computer service company, providing payroll services to third parties, including some of Norvic's competitors. "We had to set up walls to make sure that information didn't leak," Henri remembered.

This got the attention of a local East Anglian newspaper, which interviewed Henri Termeer for the first time.

"They were quite curious and I was excited," he recalled, "particularly about being able to say, 'we can service the company, but also generate revenues and actually break even as a department.' It was a valuable learning experience for running a business."

These opportunistic, entrepreneurial experiences would take root in Henri Termeer and drive curiosity about his future. He had met a number of others in the East Anglian community who had attended graduate school. These were the early days of international students attending American MBA programs, and Henri was recruited by alumni from several top schools, including the University of Virginia's Darden School of Business. Henri chose Darden, where he enrolled in graduate school in September of 1971.

While working in East Anglia, Henri had also met and become romantically involved with a younger woman, Maggie Riches, 20, a U.K. native. She too would fly across the North Atlantic, joining Termeer on his journey to America.

CHAPTER TWO

Wings

The influences that shaped Henri Termeer's development as a pioneering biotech executive began with his family life and youth in Tilburg, his service in the Dutch military, and his first job with Norvic, the U.K. shoe retailer. But it would be his graduate education at the University of Virginia's Darden Business School and his 10 years at Baxter that gave Termeer his focus, taught him leadership skills, shaped his character, and forged the "career imprints" that would eventually make history in the biotech industry.

Henri had met a Darden alumnus and well-known U.K. tennis player named John Baker while working for Norvic in East Anglia. Although he had also been recruited by a couple of Ivies, Termeer ended up making a different choice.

"John Baker convinced me to go to UVA, which was then a new school," Henri would recall. "I didn't know much about the University of Virginia. I didn't know much about Harvard either, for that matter, other than the name, and I knew nothing at all about Cornell.

"In the end, I decided to go to UVA because they expressed a real interest in me. They wanted to remedy the shortage of foreign students, so they offered me a marvelous place to live. There were only five foreign students in our class of 105, and I was one."

Darden also offered Henri Termeer an attractive financial package that included a scholarship for the first year. John Baker lived at the time in Greenwich, Connecticut, and offered to help Henri in making his way to Charlottesville.

"I had never been to the U.S.," Henri remembered. "It was a great adventure. I arrived at JFK with my girlfriend, Maggie. We had five cases—four were hers, one was mine. We sat innocently in the terminal waiting for a bus to Greenwich, Connecticut. When the bus arrived, we realized that my girlfriend's suitcases had been stolen. I had my case and she had nothing. It was very sad. We were running around for a few hours and eventually took a later bus. That was a good lesson."

Henri and Maggie regrouped.

"We stayed in Greenwich for two days, and then took a Greyhound to Charlottesville from 42nd Street. It was September, still very warm. The bus was full with all kinds of people. Those first moments in a completely new country really made an impression on me. And that bus took forever! In the Netherlands, you can get anywhere in half an hour."

Charlottesville was a small, bucolic college town in the Blue Ridge foothills of central Virginia. On arrival, Henri and Maggie collected their few belongings and settled into the living quarters Darden had chosen for him. Henri remembered his monthly rent was "something very marginal."

Their student housing turned out to be an old slave cottage, a white clapboard dependency called "Holly House," which was situated right behind Faulkner House, an historic, 100-year-old mansion that had been named for the illustrious Southern author who had spent five years in residence on the campus.

Occupying the first floor of Holly House was a storage room for the groundskeeper's tools and gardening supplies. Henri and Maggie lived upstairs. He could walk to class on Monroe Hill. The conditions were not great, but the price was right.

But there was a slight problem.

"This was the South," Henri remembered. "They gave me the house, but not to live in sin. There was frowning as soon as we arrived. At orientation, they mentioned that I would have to work day and night because the program involved three cases a day and an examination every seventh day.

"They said I wouldn't be able to look after my girlfriend, and that she could only stay six months because she didn't have a student visa. If she married me, all of that would change."

So, days after their arrival, Henri and Maggie decided to get married.

"We notified the school and our 'big brother,' a second-year student whose family was looking after us. The family organized the wedding behind our backs. They said, 'don't worry about a thing.'

"They organized a wedding on Jefferson's old grounds. It was probably one of the very few weddings to take place there. It was quite lovely. They asked us what we liked. I said, 'I like classical music.' My wife said, 'I like candlelight.'

"They put candles around a beautiful, classic garden and invited all of the students and faculty. Our parents weren't there, but they made it magnificent. Somebody gave us a car, and we got a room at the Holiday Inn in the Shenandoah Mountains for one night. We had new suitcases, of course.

"The wedding was a tremendously powerful statement of welcome. Nobody knew us, and we had only been in the country a month. Giving us that great embrace was, I thought, very impressive. That was my first impression of the U.S."

Termeer would describe his time at Darden as "lots of work and lots of fun.

"I did a lot of things on the side," he said. "They had a consulting group, and I got some credits for doing that. I did some work at the World Bank and wrote my thesis there. I audited a course on international law. I made great friends, learned a lot, and I loved the case method. All of it convinced me that this was the right direction for me."

Henri appreciated what he called "the human feel" of Darden. "It was not culture specific; it goes across boundaries. The things I learned were particularly strong on the management side, the understanding of the tremendous power of the human factor in the business process."

"Darden played a special role for me—it was my gateway into this country. I arrived with no wish to immigrate. Darden is responsible for my change of heart, providing my foundation, my base, and my first network in the U.S."

He would soon discover another network that would not only prove invaluable to his personal growth, but become nothing short of legendary in the business community.

Between his first and second years, Henri went back to Europe to look for a job.

"I went to Unilever, the consulting unit of the *Economist* magazine, and Royal Dutch Shell. They were very unsatisfactory. They didn't have a good understanding of what an American MBA meant. That experience made me want to work in the United States. UVA had an on-campus recruiting program. Eventually I ended up at Baxter Travenol."

Henri's interview with Baxter was the last one he had sat for on campus. He was told the company was looking for people who spoke European languages and understood European cultures to become general managers in Europe.

Baxter had been founded in 1931 by Dr. Ralph Falk as Don Baxter Intravenous Products, Inc., after the physician who pioneered the first intravenous solutions brought to market. In 1953, William Graham, a patent lawyer who had joined Baxter in 1945 as a vice president, succeeded Ralph Falk as CEO, a position he would hold for 27 years. He would come to be known as "Mr. Graham."

By 1973, Baxter had become a hot, rapidly growing company whose leadership ranks were swelling with high-potential young MBA superstars. In her book, *Career Imprints,* Monica Higgins, a Harvard professor, explored the factors that made Baxter a breeding ground for the most successful biotechnology ventures in the United States. She explained how Baxter's "career imprint"—the result of the company's systems, structure, strategy, and culture—stayed with employees throughout their careers, enabling them to grow into great leaders.

Higgins' book is a detailed study of the former Baxter managers later to become the famous "Baxter boys," who would come to lead the biotech industry. A disproportionate number of the CEOs of early biotech companies were former Baxter employees. More than one in five biotechs that went public between 1979 and 1996 had a "Baxter boy" on the IPO team.

Even though the Chicago-based corporation had only $270 million in net sales at the time, Termeer quickly concluded that Baxter was his best option. He was attracted to not only its growing reputation as an outstanding management training academy but also its global footprint. He thought he might go there, serve for a couple years, learn a few things, and return to Europe. He might even return to Tilburg and join his brother, Bert, in running the family shoe business.

Mr. Graham personally interviewed recruits, often by traveling to business schools to meet candidates, usually chosen from the top of their classes. Henri Termeer's first interview with Graham took place at the company's headquarters, and although he may not have known it at the time, Henri had just met his first Baxter mentor.

People would often talk about how passionate Bill Graham was about patient care, eclipsing the personal dedication and emotion of all his other employees, except one. Monica Higgins quotes a businessman who said that Graham's "only rival in his passion for saving lives is Henri Termeer. Henri can make you cry."

Henri would later emulate many of Mr. Graham's management techniques. Graham typically worked every Saturday morning, and executives who wanted to impress him would show up then too. The Saturday schedule included a regular coffee break in the company cafeteria, giving everyone the opportunity to interact with Bill Graham.

It was important to Graham that he remain visible to his employees and accessible to everyone. He was "managing by walking around" before there was such a term in common usage.

One of the most important reasons Henri decided to join Baxter was the tremendous—and rapid—opportunity for growth. Graham oversaw an in-house mentoring program that placed recent MBA graduates as assistants to managers, and within six months to a year they would be put in charge of an international operation as the company expanded. When Baxter started the program, annual revenue was $100 million. It was an $8 billion company when the program ended.

The company's initial training quickly exceeded Henri Termeer's high expectations. His first assignment, as was typical for its MBA new arrivals, was a three-month project. In Termeer's case, it was an "Assistant to" role, Assistant to the VP of International Marketing. The short stint was to focus on Chagas disease, a potential new market for Baxter.

"I didn't like being an assistant," Henri would remember, "so I asked them for a real job. After three months, I became the International Product Planning Manager—a big title for a young man."

The job focused on Baxter's blood products division, Hyland Therapeutics, an Orange County, California subsidiary. Hyland made plasma products (proteins derived from human plasma) and diagnostic products. It was the only division of Baxter not located in Chicago.

Henri Termeer came to see this period as the beginning of biotechnology.

"You took plasma and pulled it apart, fractionated it. Hyland sold Factor VIII, Factor IX, immunoglobulins, and albumin. The plasma was collected through plasmapheresis performed at collection centers all around the country. They paid people for plasma.

"At the time, I was primarily trying to figure out how to make diagnostic and therapeutic products, how to market them, how to run the sales force, and how to prepare. I wasn't a specialist on these products. I learned as much as I could."

Chagas disease remained a very important part of Baxter's operations. Chagas is a parasitic disease that is sometimes called "sleeping sickness," and it is prevalent in Latin America.

"We developed tests for Chagas disease based on feedback indicating that it would be a big market," Henri recalled. "Baxter asked me to head up this project. They said, 'Figure out a way to set up the connections.' That was a very Baxter thing to do."

It would become a Termeerian thing to do as well. Figure it out. Make the connections. It is your responsibility. And do not hesitate to be bold.

In early 1974, at the height of the U.S. oil shortage, OPEC's decision to embargo the export of oil to the United States was crimping the economy. Mr. Graham assessed Baxter's financial positioning and concluded that a corporate-wide belt tightening was in order.

Graham selected five of his young executives, one from each of Baxter's four operating units and one from corporate. Termeer, living in Costa Mesa, was selected to represent the Hyland Division.

"It was decided that we needed to squeeze assets, inventories, and accounts receivable to make operations as efficient as possible. A team of five people was pulled together. One was heading it up, reporting directly to the CEO, and for each of the four divisions there was one person pulled out of the general corporation to focus on completely reengineering how business was done."

Henri Termeer began to stand out among the new group of executives reporting directly to Mr. Graham.

"The others were young, but they had been with the company for three, four, or five years. I had been there less than a year."

The project was aimed at increasing efficiencies and reducing operating expenses, tasks that nicely aligned with Termeer's Dutch instincts and previous logistics experience.

The assignment took Termeer and his wife to Brussels for three months, leaving behind their unoccupied apartment in Costa Mesa and house in Evanston, Illinois. Frequent international travel became a normal feature of his Baxter experience. It was a peripatetic, rather intense life.

"It was such a focused project," Henri said. "Reporting directly to the CEO, you could get a lot done. It was a very good experience, but it only lasted three months.

"Afterwards, they asked me to come back to Chicago for a new position. I stopped working on the Hyland stuff and became the International Marketing Manager for the Artificial Organs Division—artificial kidneys, dialysis equipment, heart/lung machines, stuff like that.... It was very interesting. Great pioneering work was being done in dialysis at that time, and in the development of heart and lung machines for open heart surgeries."

By mid-1975, after two years in the Artificial Organs Division and approaching his 30th birthday, Termeer was eligible for a new assignment. He was asked if he wanted to become the general manager of a joint venture in South Africa.

"I spent three weeks there, but I didn't like it. Their political circumstances were very complex. Plus, there's a long history with the Dutch in South Africa, so I decided against it."

He thought he was going to end up in Brazil, but Henri Termeer had caught the attention of Baxter's head of Europe, Gabe Schmergel.

Gabriel Schmergel, a Hungarian-born engineer known for his swift ascendancy up the Baxter chain of command, was a formidable, talented leader and clearly one of Mr. Graham's top proteges. He had joined Baxter in 1967 out of Harvard Business School (HBS), graduating as a Baker Scholar. Within two years of joining the company, he had been named General Manager of Baxter's important German subsidiary, Baxter Deutschland GmbH. In late 1975, Schmergel was posted in Brussels, responsible for Baxter's sprawling European region and its aggressive team of talent.

In Germany, Baxter employed around 100 people, and it was Baxter's second largest business in the world. It was a critically important country in the international build-out. It was driving more than 20%, sometimes 35%, annual sales growth for the corporation. Facing a host of challenges, not the least of which was a labor strike, Schmergel was confronted with the reality that the time was right for a new leader to take the helm of the German business. He had run Germany for two years (1969–1971), so he knew how tough a place it could be. The unions were not cooperating, the competition in the German marketplace was fierce, and the stakes were high.

Schmergel had been the head of that five-person executive committee Mr. Graham had charged with freeing up working capital, and he remembered "this young Dutchman who had impressed me, and I thought 'well, he's Dutch. He probably speaks some German, or he can learn it very quickly.' I chatted with him very briefly among my trips back to the U.S. ... to see if he would show any interest, and he said 'Absolutely, yes.'"

Schmergel conferred with Mr. Graham and his boss, head of International Bill Gantz. They agreed. Termeer was their pick. It would be a huge step-up for the 29-year old prodigy. It was, in the Baxter parlance, Termeer's "sink or swim" moment in the making. It was also one of the shortest transition processes ever witnessed.

"I had arrived back from a business trip in Europe on a Saturday and on Monday morning Gabe Schmergel called asking 'How was your trip?' I had visited one country a day, and when he asked about Germany, I said 'Well, they have a lot to learn, but they're getting there.'

"Then he asked, 'how would you like to be general manager, the *geschäftsführer*?' I was shocked. The general manager had gotten himself in trouble with the unions and there was a strike. Baxter was very anti-union. They were concerned. They made a decision to change the management.

"I called my wife and asked, 'what do you think?' That night, I took a red-eye back to Europe. I went to Brussels and picked up my new boss. We flew to Munich late that night. He met with the incumbent general manager in the airport, while I stood behind a pillar so he wouldn't see I was there. They had dinner, and he resigned."

The next morning at 7:00 a.m., Schmergel installed Termeer as Baxter's General Manager of Germany. Termeer spent the first few hours that day meeting his new colleagues, walking the halls, shaking hands, introducing himself. He could speak German, which of course was essential given what he faced.

Henri Termeer would often talk about his first night as general manager in Germany.

"I didn't eat. I drank some wine, and then ... it hit me. 'Here I'm going to run this; this is me.' It was a magnificent feeling—[that] somehow, Gabe and Bill Gantz and others thought this was an acceptable risk, to send this young guy and to introduce him for an hour or two, leave, and then let him do it. And it worked.

"Of course, once it started to work, there's this great experience in terms of your confidence building, you start to think, 'Yeah, this is actually OK. I am OK for this. I can do this. And it makes sense, what I'm doing here.'"

Henri Termeer was going to need every ounce of optimism and confidence he could muster. By 10 o'clock of his first day on the job, all of the employees except for the management had gone on strike. Termeer recollected, "Big signs appeared and women came in wearing black as if in mourning."

As the weeks went by, he won over the workforce, and during his three years in Munich, he would more than double the net sales of Baxter Deutschland GmbH and treble its work force to more than 300 full-time employees.

The "Baxter way" was akin to stretching a rubber band. Take a hugely talented, relatively inexperienced leader/manager and place him (nearly all of the MBAs hired in this era were men) in a role beyond his experience curve. At Genzyme, Termeer often employed this technique as a way of identifying his top performers, developing these executives, building out his management ranks, and, in the end, providing for leadership succession.

Alan Barrell, a contemporary of Termeer and leader of Baxter's U.K. business in the late '70s/early '80s, explained, "The culture of Baxter was international expansion; the culture of Baxter was also exploring healthcare, getting into new areas, home care, innovation, technology development. It was very, very strong in renal medicine, intravenous therapy, and a whole load of areas where technology combined with the thinking about healthcare to make sense for the patient.

"Baxter was a real pioneer in bringing new technology to market in a way that saved so many thousands of lives around the world. We used to say Baxter was the university of the healthcare industry because we educated so many people who often went off and formed other companies. Henri's apprenticeship in Baxter was what really brought him to the position of being ready for what he did with Genzyme."

Termeer's personal life was buffeted over the course of his Baxter years, as he relocated and traveled from country to country, sacrificing local ties, placing strain on his marriage, but demonstrating his allegiance to the Baxter mission of becoming one of the world's top healthcare products companies.

Later in his career, at Genzyme, Henri Termeer would reject the competitive culture that was so ingrained at Baxter.

"Baxter ... did not have a very well-developed human culture," he told Monica Higgins. "The human factor in Baxter in terms of the interaction between people, and how they worked with each other, was much more competitive than some environments... . And I put enormous value on the human factor."

As Henri Termeer took on more challenging management roles, he was becoming known as an imaginative, well-liked, grounded, and trustworthy leader. At the same time, he could be tough and at times even stubborn. But most of all, he was a leader who enabled others around him to be successful. He fed off their success. He did not put people in a box and was constructive in his criticism. He was deemed to be controlled, plugged in, and fair. He listened. And his innate ability to relate to people at all levels of the organization became legendary. Furthermore, he demonstrated an almost photographic memory for remembering names.

Beyond his clearly outstanding cultural acclimatization and leadership development, Termeer gathered two other important lessons through his German experience. Both derived from the marketplace; both involved rare genetic diseases.

Thirty-five years before Termeer had come on the scene to lead Baxter in Germany, the Nazis had controlled the country, and with them, certain stigmas were associated with various parts of its population. Under Hitler's regime, one such group was citizens with genetic diseases.

"I learned a lot about the history of rare diseases in Germany, and how they were viewed ethically," Henri Termeer would remember. "In the Second World War, citizens with genetic diseases were in a very difficult position. They were ostracized, not supported at all. Many families went into hiding for generations. I learned a lot about the international politics of genetic diseases. These were life lessons that are still valuable to me today."

Another lesson Henri Termeer learned pertained to the registration, marketing, sale, and pricing of prescription pharmaceuticals that treated rare genetic diseases. One such disease found among Germans, and most common to males, was hemophilia. Germany was the largest single market Baxter had for Factor VIII, the coagulant factor used to treat it.

Henri Termeer would be exposed to a way of thinking, and an organizing principle for society, that would prove hugely influential.

"A law in Germany at the time said something like 'No citizen, if it can be avoided, should be disabled.' If you can manage the health of a citizen in order to avoid an ailment, then that individual has an absolute right to treatment. In hemophila, that means stopping the bleeding that might otherwise occur. Germany was very decisive in the way they looked at the treatment of hemophilia.

"Of course, there was a global shortage of product that drove the price up. The German prices were higher than elsewhere, but their consumption was also higher. Other countries would complain that the Germans were cornering the market. The ethics of the business were very interesting."

Bob Carpenter, one of Termeer's Baxter colleagues and later Genzyme's lead independent Board Director, saw how Henri Termeer's immersion in patient care informed his decisions as a manager. He had learned that some physicians were treating patients with serious blood coagulation disorders who had developed inhibitors of Factor VIII.

"Normal dosages of Factor VIII didn't work," Carpenter recalls. "There was a doctor who decided he wanted to try to help these kids by giving them massive doses of Factor VIII. At the time, Factor VIII sold for four cents to 10 cents a unit, but you needed hundreds of thousands of units. This guy gave millions of units to these kids and it worked."

Termeer tuned into the impact of dosing patients, regulating the dose to modulate its biological response.

Carpenter continued, "Henri worked with a physician in the Bonn Hemophilia Center who treated pediatric patients who were entering surgery. Many were young boys with hemophilia. He also worked with the government. Henri got them to realize that massive amounts of Factor VIII, sometimes costing nearly a million dollars, would save their lives. And he also got them comfortable with spending these sums, particularly in a rare disease situation, where it's a relatively small budget item for the government."

By the end of 1979, at the age of 33, Termeer's tour of duty in Munich had been completed. He was ready for bigger challenges. Bill Gantz called and asked if he would move to California to take up a new position, returning to their blood products division, Hyland, as executive vice president. The role was expected to lead eventually to him stepping into the position of President.

He would report to Dave Castaldi, the incumbent Division President, and he would assume responsibility for its R&D, global marketing, and regulatory affairs groups. Hyland was the first of Baxter's units to globalize. Having served in Germany where blood products were so central to the business, Termeer would hit the ground running. He would be based in Glendale, California.

Bill Gantz described Termeer and his fit with the assignment, "We asked him to go out to Hyland, where we had some challenges. This again was typical of Baxter. We moved people around to give them experience, but they also had to be willing to take the challenge... . We didn't move them into easy spots.

"Henri had a nice personality. He was not only bright, he had good judgment ... and he was enthusiastic and infectious... . He had this ability to communicate with people... . When you're in international you understand how important communication is because it's far more than just language. You really have to have people who are able to engage and understand the cultures and understand differences, and Henri was one of those... . He was one of the stars of the group at Baxter."

Termeer settled that winter in Glendale to take up his new post. It was his first experience in managing research and development. Later, he would add manufacturing to his portfolio. Germany's organization had been limited in scope to sales, marketing, and regulatory affairs. Years later, this expansion of functional experience would prove pertinent to his success at Genzyme.

Termeer was not formally educated in scientific matters, but by now, he was well versed in the biology of whole blood, plasma, platelets, and the fractionation processes that led to the creation of pharmaceutical products derived from human material. He was deeply aware of the peril of contaminants and the absolute requirement that Hyland's products be manufactured and finished in an immaculate, pure, unadulterated condition. "Particulate matter" would contaminate the product and was simply not permitted by the FDA or, for that matter, by physicians or their patients. The finished, vialled product had to be perfect. Every time.

For the next four years, Termeer reported to Castaldi, who had been assigned to lead the division in very late 1977. Castaldi, also a Baker Scholar graduate of HBS, was among the brightest, highest-potential leaders who had come up through Mr. Graham's MBA program. Stylistically, the two executives had different tendencies; one was outgoing and the other more reserved. Although certainly not in a ruthless way, both were ambitious. They were nearly the same age.

Castaldi remembered Termeer as someone who would go on to become "the greatest CEO in the history of biotechnology."

Later in his career, Termeer drew from a Baxter experience, that of introducing a new product, a second generation of Autoplex-T, to the U.S. market. It was a drug therapy used by hemophilic patients with Factor VIII inhibitors, just like those he had seen in Germany. Autoplex-T was one of the industry's first value-priced products. Its launch would become directly applicable to Genzyme's 1991 introduction of Ceredase for Gaucher disease.

About 10% of hemophilia A patients, an estimated 1,000 people, suffer from Factor VIII inhibitors. It is an ultrarare condition. When Baxter initially introduced Autoplex-T, outside the U.S. market, the annual patient cost approximated $50,000, a staggering price. The product was in certain instances, lifesaving, and the demand for it was strong. It became among Baxter's most important products in Germany and among Hyland's most important hemophilic products globally.

Termeer learned immensely from the launch of Autoplex-T. He learned the importance of establishing relationships with the regulatory authorities; they were responsible for approving the product. He learned the value of a lifesaving medicine. He learned the importance of getting close to the patient, placing him or her at the center of not only the delivery of care but the clinical development of the product. He acknowledged and facilitated the critical role of physicians. He also proved that payers, be they from the public or private sector, would reimburse for it.

The Autoplex-T experience was enormously instructive and provided Termeer with precious insights he could have gotten at few other places. It would prepare him for what was to follow. Because of its high cost, patients and payers "would scream at us," as Bob Carpenter relayed. But the lives of rare disease patients were ultimately saved. Henri came to realize that "people would pay." The development and introduction of Genzyme's rare disease franchise products in Gaucher disease, Pompe disease, Fabry disease, and mucopolysaccharidosis type 1 (MPS 1) would later be directly impacted.

Henri Termeer's return to Hyland would leave an indelible mark, as a future entrepreneur, on his understanding of the potential of treating rare disease patients.

"It was an extremely challenging environment, but very nice," Henri would later say. "I was in my early 30s by then. American industry was moving rapidly. I couldn't see stale European firms making the same kind of progress. I was enthralled by American entrepreneurship. Baxter was very entrepreneurial. There was tremendous camaraderie in the company. People got to do things they had never imagined."

The training and development Termeer received at Baxter stands out as the most important of his formative career experiences. He would infuse Genzyme with many of Baxter's cultural values and operational methods. From those of his generation at Baxter, Termeer's legacy would go down as among the most true to the company's imprint. It would ultimately redound to Genzyme's benefit.

CHAPTER THREE

A Great Convergence

As 1982 drew to a close, Henri Termeer's thoughts turned to potential new challenges. He had been at Baxter nearly 10 years, and his career's ascent had slowed if not stalled. Uncertainty around his trajectory was creating at least some degree of anxiety. The life sciences arena was turning to genes, a direction Baxter was slow to embrace. As the New Year was dawning, so too was a new age.

Baxter's commitment to biotechnology was faltering under Mr. Graham's recently named successor. "Vern Loucks took over from Graham and the company became much more device-oriented and less science-oriented," Termeer remembered about the transition. "Biotech was just beginning... . The company had been told by a consulting firm that it was too late for Baxter to develop biotechnology capabilities because the established companies like Biogen, Genetics Institute, Genentech, and Amgen had already built an experience curve that couldn't be matched.

"It was a well-known consulting firm—bad advice," Termeer later chuckled.

Since 1953, the year in which Nobelists Jim Watson and Sir Francis Crick had solved the double-helical structure of DNA, the world had been waiting for the ensuing breakthroughs that would transform medicine. It had been a long wait, but by the late 1970s, industry had finally caught on. Biotechnology companies began to sprout, largely through the works and ambitions of scientific founders affiliated with major research universities. Indeed, Genentech, Biogen, Hybritech, Centocor, Genetics Institute, and Amgen had all been founded in the five years 1976–1980. They were situated in one of three primary geographic clusters that would dominate the early years of biotechnology: the San Francisco Bay Area, Southern California, and Boston/Cambridge. All were venture capital–backed, possessed an all-star Scientific Advisory Board, and were led by entrepreneurial founders.

For Genzyme, however, which was founded a year later in 1981, the similarities to this cohort would stop there. Genzyme would prove to be a different kind of biotech enterprise with a unique, niche strategy and a leader who marched to the beat of a different drum.

Genzyme grew initially out of the work of a young enzymologist who, as it turned out, had never graduated from college. Henry Blair was a resourceful, ingenious researcher, running day-to-day operations at Tufts Medical's New England Enzyme Center. Responding to a request for proposal (RFP), Blair had won a contract with the lab of NIH's Roscoe Brady to produce the enzyme needed for Dr. Brady's work in placental-derived enzyme replacement therapy. Among other things, as a means of securing material for Brady, Blair's brother drove around Boston and collected placentas from local hospitals. It was the only tissue that had a sufficient concentration of the enzyme Brady needed.

"I got to know Roscoe very well," Henry Blair recalled as he reflected on Roscoe Brady and his influence in those early days. "He became a mentor to me, and I learned an enormous amount of science from him. Roscoe was an amazing individual—one of the few people who discovered the defect in a disease and went on to come up with essentially a cure."

As Blair was looking to find ways to expand his work beyond what NIH would fund, Sheridan Snyder, a serial entrepreneur who had previously realized handsome returns buying and selling two packaging companies, sought his next gig.

Snyder, a French major and 1958 graduate of the University of Virginia, approached his old friend Ed Glassmeyer, a former Donaldson Lufkin & Jenrette (DLJ) venture capitalist. In the late 1970s, Glassmeyer had left DLJ to form his own venture capital (VC) firm, Oak Investment Partners. A long sought change in the Employee Retirement Income Security Act of 1974 (ERISA) pension law had been enacted in Washington. The event had thrown down the gauntlet for professional investors to start high-risk institutional investor funds. Venture capital firms were among the major beneficiaries, and the floodgates of venture investing were opening.

Glassmeyer tapped one of his associates, Eileen "Ginger" More, to work with Snyder to sponsor Oak's investment in his new enterprise, whatever that would turn out to be.

More had been on track to graduate from the University of Bridgeport, Connecticut as a math major. Oak had hired her because she had obtained a Chartered Financial Analyst (CFA) certification while working for Wright Investors' Service. More recalled Glassmeyer's rationale in hiring her, "'We need someone to look at deals and run modeling.'"

Weeks later, on the hunt, Snyder and More were tipped by one of Oak's limited partners, 3M Corporation, that the technology being developed by Blair might serve as the basis for a new start-up. More had as much experience in evaluating life sciences investments as did Snyder, meaning little to none. The technology was focused on the use of enzymes derived from placentas to treat some obscure disease, something called Gaucher disease.

"They're doing some interesting things," More was told. "One of the things they're doing is a penicillin assay for cow's milk so that you can measure how much penicillin is in the cow's milk, which is important because we don't want to feed our kids penicillin in milk."

Although More found this idea potentially backable, it did not really capture her imagination. Another one did.

She remembers the 3M tipper's words, "We have one other scientist, Henry Blair, who is working on a disease called Gaucher disease, and he is extracting an enzyme from placentas. We're not doing this kind of thing, but we think perhaps there is something there."

How More and Snyder figured out that the makings of a great biotech company were to be found at Tufts' New England Enzyme Center, harvesting placentas, would prove to be one of the greatest mysteries, and successes, of bioventure investing ever.

But there it was, a platform investment that would begin with Blair, the NIH contract, and the placentamobile. Snyder would also roll up two unexciting but financially strategic lab products businesses based in the United Kingdom, Whatman Chemical and Koch Light, thereby forming the nucleus of the early version of Genzyme.

It was More's first deal, and it would prove to be pure genius. Oak had paid four cents a share for its initial seed stake.

A few months later, John Littlechild, and his firm, Advent International, would acquire Series A shares in Genzyme. As a part of its deal, Littlechild would join Snyder and More on the Board.

Around them, the biotech world was on fire. Genentech, the pacesetter, had gone public October 14, 1980. Its shares had nearly tripled on the opening of trading, soaring from $35 to $89, closing at $56 by day's end. Genentech, whose stock symbol was DNA, had been the first to catch the wave of first-generation biotech IPOs that would run from the early 1980s into late 1987. Its founder biologist, Herb Boyer, would appear on the March 9, 1981 cover of *Time* magazine heralding a new age, "Shaping Life in the Lab: The Boom in Genetic Engineering."

Wall Street was dazzled ... and abuzz. Biotechnology had entered the mainstream and was destined, as the cover story propounded, "to be the technology of the 1980s, just as plastics were in the 1940s, transistors in the 1950s, computers in the 1960s, and microcomputers in the 1970s."

Over the course of the next few quarters, Genentech would be followed by Cetus, Hybritech, Centocor, and Amgen as newly public companies. One of these IPOs, Hybritech's, was led in 1981 by one of Termeer's former Baxter colleagues, Ted Greene, who was 41 at the time. Greene often referenced Baxter and his years at the company as part of his impetus in starting a biotech business. Prior to working for Baxter, he had also been at McKinsey and consulted for Mr. Graham. He regarded Graham as "the most accomplished, tremendous CEO I've ever known or worked with. He had a very interesting philosophy in management and that was he simply hired the best and the brightest and the most ambitious people that he could hire and then created something for them to do."

It seems as if Greene and Termeer would eventually read off the same sheet of music. So would many others. Of the 299 U.S. biotech companies that would go public from 1979 to 1996, more than one-fifth of the CEOs had spent time at Baxter prior to joining their new company.

Kleiner Perkins Caufield & Byers (KPCB), the Silicon Valley VC firm, had founded Hybritech and attracted Greene. They had also recently backed Steve Jobs in launching Apple Computer. Brook Byers, a junior KPCB partner, had invested $300,000 in Hybritech and rented lab space. The second California gold rush was on. Four of the first five major biotechs were based in the state. And investors were swinging for the fences.

Greene later captured the zeitgeist of the age, "The mood was like a game. We were trying to solve a mystery of what can be done... . We were on the cutting edge of science, and that was the spiritual glue that held us together."

Meanwhile, Genzyme was searching for some spiritual glue of its own. Its connection to cutting-edge science could hardly be found in Whatman Chemical, Koch Light, and harvesting human placentas.

But Sherry Snyder had an idea. He thought he might find the missing ingredient at MIT. He called noted molecular biologist, Harvey Lodish, PhD, to explore an idea, that of linking in some, to-be-determined way with one or more faculty members to create Genzyme's initial scientific advisory board.

In the very early 1980s, faculty members of leading universities were being approached from every direction by hungry venture capitalists who were seeking to found new companies and leverage affiliations with the most prominent names in science and medicine. Everyone could see the explosive growth ahead, and, among other things, talent and connections would serve to differentiate the biotech field.

Around the time of Snyder's outreach, it happened that Lodish, six other MIT professors, and one from Harvard had hatched an idea of their own. To address the avalanche of inquiries they were receiving, they were going to form a consulting company, BioInformation Associates, Inc. (BIA), and organize themselves to assist their clients in evaluating technology and their potential for commercial exploitation. The partnership's eight scientists offered world-class expertise in a constellation of life sciences disciplines—from molecular biology to bioprocessing, chemistry, microbiology, materials science, and enzymology.

Their multidisciplinary approach could handle almost any scientific biotech question on the planet. This group of eight rivaled that which any biotech company could assemble anywhere, and in terms of prestige, credibility, and relevancy, it was a Genzyme home run.

Formalized in April 1983, Genzyme would enter into an advisory relationship with BIA. Among other things, their agreement provided Genzyme with a ten-year exclusive, worldwide license from MIT to proprietary technology in remodeling glycoproteins, including what became Ceredase. It also gave BIA a 10% equity stake in Genzyme.

The relationship would also create in one fell swoop Genzyme's scientific advisory board. One BIA founding principal, Charlie Cooney, PhD, an MIT expert in bioprocessing, would later describe its fit within the overall Genzyme portfolio as the company's "fourth asset." Scientifically, Genzyme was now connected to the best of the best.

To represent BIA and its owners, Cooney would join Genzyme's Board of Directors, a directorship he would hold until 2011.

Short of later hiring Henri Termeer himself, Snyder's most significant contribution in cofounding Genzyme was the creation of this linkage with BIA.

While this was taking place, Henri Termeer kept his head down as the headlines were lit up with stories of biotech start-ups, IPOs, the next big cure, and discoveries pushing the edge of medicine. Nonetheless, Henri's phone began to ring, and the siren song of leading new ventures with it.

In early 1983, after many months of declining to explore the biotech opportunities that were being trolled in front of him, Termeer eventually decided the time had come to look outside Baxter and evaluate some of them more seriously. He was under no pressure to leave Baxter, and his compensation was substantial as the second-in-command at Hyland. But the handwriting was on the wall. The future was now, and now was his future.

In parallel, it so happened that Ginger More, John Littlechild, and Sherry Snyder concluded that the time had come to hire into Genzyme a strong operating executive with a highly pedigreed background in pharmaceutical products. The goal was to find someone who could take Genzyme to a new level and provide for an orderly CEO succession down the road.

"We needed to bring in somebody who could really bring about some element of structure if Genzyme was going to evolve," Littlechild would remember. "The key player here was Sherry. Whoever came in had to be acceptable to Sherry... ."

Weeks later, Ginger More was having dinner in New York with Fred Adler, a successful venture capitalist who had encountered an impressive executive, Henri Termeer. Adler suggested they ought to consider Termeer for the role.

Termeer had nearly been selected as CEO of one of Adler's portfolio companies, Life Technologies, based in the Washington, DC suburbs. They made sera and other products that served as the picks and shovels of biotechnology labs. Adler's search had just closed with the selection of another candidate. The Board had liked Henri, but had chosen another executive to lead the business. Termeer, therefore, remained available for consideration elsewhere.

To determine his interest in a leadership role with Genzyme, More and Snyder would take the initiative and directly approach Termeer in the summer of 1983.

Soon after, Gabe Schmergel's phone rang. It was Henri, "Hey, Gabe. I'm going to be in Boston next week, and I want to see you."

Termeer was looking at a new opportunity and wanted his take. Although they had not remained in close contact, they always got along. And Henri had taken note when Gabe left Baxter in 1981 to pursue a CEO post of his own, joining Genetics Institute (GI), a hot Boston biotech start-up backed by Greylock. Its two founding scientists, Tom Maniatis and Mark Ptashne, were Harvard molecular biologists and rising stars. Henri was in Boston the next week. He and Schmergel would get together.

Termeer entered the GI office that summer morning. It was located in a sketchy neighborhood, Mission Hill, in an abandoned, large, brick building. It had been the Boston Lying-In Hospital, one of America's first maternity hospitals.

A watchman greeted Termeer. At the time, GI was the building's only tenant. In Schmergel's own words, "It was creepy because here's this huge facility, and we are the only people there. I'm sitting in my office when Henri walks in. I had a very small desk and a telephone. That's it.

"I remember Henri is looking at me and is, oh my gosh, I can see it in his eyes. 'My god, this guy used to be in charge of 10,000 people at Baxter, and he's now sitting here in a forlorn, forgotten corner of this empty building.'"

Termeer opened the conversation, "Well, Gabe, what's your job here? What do you do?" Schmergel grinned and replied, "Well, I'm trying to manage a bunch of unruly, brilliant scientists. And I'm trying to raise money." And then Termeer asked, "Do you like it?" to which Schmergel grinned again and replied, "Henri, I love it."

At this point, Termeer's face lit up. He had expected to hear about Gabe's "big mistake."

But the conversation with Schmergel was not over. Prior to the meeting, he had sent Schmergel a briefing package on Genzyme. Schmergel had looked it over and asked his scientists what they knew about Genzyme. His assessment was rather grim, "Henri, this looks like a very pedestrian kind of opportunity."

About the same time, Termeer sought another sounding on the attractiveness of the Genzyme opportunity. This one came from Jim Geraghty, a young consultant at Boston's respected Bain & Co. The men had become business confidantes over the previous two years as they had worked through a host of strategic and operational issues pertaining to the launch and pricing of various hemophilia products at Hyland.

Geraghty was in Glendale that day, visiting as he routinely did twice a month. Termeer pulled him aside and explained that he was confidentially considering an opportunity with a biotech company in Boston. He asked Geraghty to take a look at a package of materials and offer his advice.

Geraghty took a few days and was not so sure. "I told him that I thought the company was very small. The diagnostic enzymes business didn't seem very exciting. This wasn't going to be a great business. And the other thing was Genzyme didn't have a strong capability in molecular biology or recombinant engineering, which is basically what biotech was in those days. Genentech and Amgen had it; Biogen and Genetics Institute had it. Genzyme didn't. It wasn't nearly as exciting as other companies that had Nobel laureates on their board."

The conversation extended from there. Geraghty dived into the company's business plans and, most specifically, its plan to develop the therapeutic enzyme Brady was testing in Gaucher patients. He judged that this was perhaps the most exciting among the company's assets. But he calculated that to generate enough revenue to attract investors, Genzyme would have to charge up to $250,000 per Gaucher patient per year, maybe more, and he told Termeer, "That's ridiculous, nobody's ever gonna pay that much!"

It turned out that Termeer's assessment of pricing was aligned with Geraghty's, but he was firm that this was far from ridiculous, "The question is whether societies want these very sick patients treated. I think they do, and that if we explain why a therapy for such a rare disease is so expensive, they'll be willing to pay what it costs."

They also talked through the regulatory environment. The FDA in those days specified a very rigid drug approval process. As Geraghty understood it, he explained to Termeer that they would require "two well-controlled, blinded, large clinical trials... . How could you ever do two trials like that in this small, rare disease population?"

Termeer's answer was, "The FDA will change. It should change. If this is the right thing for society, if this is the right thing for patients ... then there should be a more streamlined basis for approval. We'll just have to get all the people, all the participants in the system lined up ... the patients and the physicians, and explain it to the FDA."

Geraghty remembers laughing, perhaps to relieve the tension in the room. He was unpersuaded, "Well, Henri, I think you're crazy." A few years later he would end up joining Termeer at Genzyme, and stay there for more than 20 years.

Indeed, in the climate of the early 1980s, Termeer's hypotheses were preposterous. Unthinkable. Laughable to nearly everyone but Henri, whose head, and perhaps his heart, said he would find a way.

So the dance would continue, at least for now. Henri continued to do his due diligence, and the Genzyme board deliberated, not exactly sure that Henri would be their choice as they surveyed the available talent.

The selection of Termeer as Genzyme's first President, however, would not linger. His and their decisions appear to have been reached one early fall Saturday morning on the campus of MIT in Cambridge. The BIA board had heard about this fellow, Henri Termeer, from Sherry Snyder. BIA decided to host Termeer in the Chemistry Department conference room, just down the hall from BIA partner George Whitesides' office.

What started as an evaluative interview quickly morphed into a full-blown sales pitch. Charlie Cooney too attended the session and remembered it this way, "It was to be kind of a pre-strategy session with Henri. The purpose was twofold. One was for us to get a sense of Henri and what he would bring, and, two, it was to sell him on Genzyme and give him a perspective of where we collectively felt the science was going to go in terms of products. And that maybe was magic.

"It very quickly became a love fest. It was not a 'let's evaluate how he's going to do.' It very quickly became 'we gotta sell this venture to this guy' ... his energy, his enthusiasm, his vision right from the moment he walked in the door was just invigorating. I mean the energy in the whole room went up a huge notch." Several hours later, by the time of its conclusion, Termeer had seen and heard about all he needed to reach his decision. So had they.

The wheels turned briskly, and the deal was done. An offer was sculpted. Termeer would take a substantial cut in pay, half of what Baxter was paying him, but, in return, he would get a sizable equity interest in Genzyme. He would be named President. He would report for work in October, 1983. He would relocate from Los Angeles to Boston, eventually taking up residence in Wellesley.

When Henri Termeer began work at Genzyme, the company had 17 employees in the company's Boston headquarters at 75 Kneeland Street.

It was located on the top floor of a rundown, 15-story garment district building built in 1930. Overrun by drugs and prostitution, the area around it had become known as the "Combat Zone." Prior to Genzyme's occupancy, its 15th floor office had been occupied "by a clothing company with a huge space of racks with hangers on them. No air conditioning, windows wide open to cool the place off, hot as hell," recalls Henry Blair.

"It was, of course, not much," Henri Termeer recollected from the initial impressions he drew that first week he set foot at Genzyme. "It had the right ingredients. I liked that the direction hadn't really been established. It was an open book. I could take it where I wanted it to go. It was a shot in the dark, and it was magnificent."

Little did Genzyme's board, let alone Termeer, know what the future would hold in store. "We had no idea, no clear vision as to what Genzyme was ever going to be when it grew up," mused John Littlechild. "The notion of this becoming a rare disease company didn't emerge till the 1990s."

But as the decade of the 1980s unfurled before them, the timing of Termeer's arrival and Genzyme's creation could hardly have been more propitious. A sequence of events would cascade, changing the external environment for Genzyme and Termeer, and ultimately open the door for what would become unimaginable success.

The first of these many events had occurred on January 1, 1983 when the Orphan Drug Act (ODA) became law after approval by Congress and the signature of President Ronald Reagan.

This landmark legislation created a category of "orphan drugs," defined as those medicines that would treat ailments suffered by fewer than 200,000 patients in the United States. Together with a later amendment, it also created financial incentives to spearhead the innovation and investment to develop medicines targeted at treating rare diseases. The arcane, amended decree would prove to be a big deal, the catalyst for a movement of rare disease innovation.

Companies would now be granted seven years of market "exclusivity" upon the approval of an orphan drug, meaning that no other company could market a competing drug during that period. The amendment also offered large tax incentives, enabling the innovating company to deduct half of the cost of the clinical trials to develop the orphan drug product. Although obscure, these legislative developments nonetheless would attract not only patients' but an industry's attention.

When the amendment became effective on January 1, 1984, Termeer had been at Genzyme all of about 75 days.

The ODA's impact was commonly measured by the avalanche of new rare disease therapies that were developed after its passage. On this measure alone, it was a resounding success.

A little known Connecticut mother, Abbey Meyers, whose son had a rare genetic disorder, Tourette syndrome, had spearheaded the long, hard-fought battle to incentivize pharmaceutical companies to invest in and develop drug therapies for such maladies.

Her ally had been the Orphan Drug Act's author, Congressman Henry Waxman, a young legislator from California who would go on to make a career out of excoriating the big research-based pharmaceutical companies.

With Waxman's encouragement, Meyers had confronted the industry, posing a very simple question to which this new legislation would have only a partial answer, "How would you feel if they (rare disease patients) were diagnosed with a serious disease and you couldn't get a treatment for them, not because there was no known treatment, but because no company felt it would be profitable enough to manufacture?"

Meyers relentlessly pursued newspaper, TV, and magazine coverage, asserting that the pharmaceutical industry lacked empathy and was unresponsive to a mother's call for help. She mobilized an enormous undercurrent of activism, and ultimately organized it, founding the National Organization for Rare Disorders (NORD) in May 1983.

Meyers even prompted the producers of a popular TV show, *Quincy, M.E.,* to write an episode about a teenager with Tourette syndrome as a vehicle to educate the public about rare diseases and the forgotten patients for whom there was little hope or treatment.

The show's star, Jack Klugman, had read about Meyers' Congressional testimony, and her story had resonated with him. Klugman ended up doing two shows—one that focused on Tourette syndrome and another about a neurologic seizure disorder called postanoxic myoclonus—that served to create awareness of the rare disease community's struggles.

"People by the thousands wrote letters to Jack Klugman about this show," Abbey Meyers said. Some said they knew a Tourette patient, others said they had another rare disease that didn't have a treatment on the market. Jack Klugman had told the post office, "Don't even unpack your sacks of mail, just send them all to Abbey Meyers."

A team of volunteers helped Meyers sort through the mail, keeping the Tourette letters and forwarding the rest to other rare disease support groups.

"There is no question in my mind that these TV shows, watched by millions of Americans, were the primary instrument that moved the Orphan Drug Act over the finish line," Abbey Meyers says today.

Meyers and Termeer became well acquainted over the ensuing years. They often disagreed on public health policy. But passion for patients, especially rare disease patients, was something on which they could always agree.

The rub between them would normally occur when the topic turned to drug pricing. The crux of the argument against drug makers was unmistakably clear. Notwithstanding the success that the original ODA and its amendments stimulated in the development and introduction of orphan drugs, the pharmaceutical and biotech companies were criticized for their failure to contain the high prices of the innovative drugs that followed.

The legislators had not countenanced an unintended consequence of the legislation that had been crafted, in part, in an ambiguous manner. It would allow some unscrupulous drug makers to abuse the law by sharply raising prices on old, existing drugs while making little or no R&D investment in these products, and thereby adding little or no value to their therapeutic profiles.

Despite some challenges in its implementation, the Orphan Drug Act would be hailed as a monumental achievement, stimulating new therapy development and empowering rare disease patients, their families, and caregivers. The meteoric rise in patient advocacy is commonly cited by healthcare professionals as having a direct link to the Orphan Drug Act's passage.

It had shifted the playing field in favor of these groups, which for decades had suffered neglect or worse. Henry Waxman proclaimed at the time of its passage that "we have reordered the economics of pharmaceuticals to make the market work." After all, many of those who had engaged in the politics of the ODA's creation would later concede that its market-based approach had not been a choice, but rather a necessity.

NORD was one of the first to raise the flag for orphan disease patients, but the wave was just starting as hundreds of other patient advocacy organizations would within a decade provide the clarion call for the creation of patient registries, payer advocacy, access to new medicines, and reimbursement. Genzyme and Termeer would directly benefit from and engage with this passionate community.

The next factor lifting Genzyme's wings was the concern for acquired immunodeficiency syndrome (AIDS) and the deepening threat it would pose to not only the nation's blood supply, but to all medicines derived from human sources—medicines such as those derived from the fractionation of whole blood and human placentas. As Genzyme was developing its first rare disease product, Ceredase, this international crisis was leading to an outcry from clinicians and patients who would potentially suffer from contaminated drug product.

In mid-1986, AIDS-related complex, or ARC, was designated a rare disease by the FDA, and with its designation, the focus on rare diseases grew exponentially. The awareness would call attention to the challenges faced by small patient populations. The AIDS community's impassioned and vocal calls for help from the biotech industry served to enhance the visibility of the most vulnerable of patients—those for whom no therapies were available.

The ARC crisis also brought to the fore the importance of biological manufacturing methods. All of a sudden, the purity of the very sources of certain drug products, human-derived material, was being threatened. Recombinant DNA technology held out the hope, if not yet the assurance, that it could deliver a solution. Genzyme, as a company, was developing its technological methods for recombinant DNA (rDNA) production in the thick of the raging debate over what to do to protect patients.

During this period, other external factors would generate wind at Genzyme's back. The global equity markets were lifting stock prices around the world. If ever there was a mother's milk for biotechnology, it was easy money. And where else could the money have been easier than from the equity markets in the mid-1980s? During the five-year-long U.S. bull market for stocks, which began in August 1982, the U.S. Dow Jones Index soared 250% until its crash on Black Monday, October 19, 1987. The market, however, recovered the entirety of its losses in less than two years, and the industry's lift was renewed.

The flow of capital into biotech, from public and private sources, was undoubtedly a major contributor to the robust growth and excitement that was enjoyed by the companies engaged in the new recombinant genetics age. Investors, employees, physicians, and, of course, patients would all benefit and prosper from the advances made in these generous years.

With this sweep of history and momentum, Termeer and Genzyme would have near-ideal conditions to tackle the building of a major, important, global healthcare products company, anchored by a franchise in rare disease therapies and diversified by business segments in other less risky healthcare product niches.

As he took up its Presidency in 1983, Henri Termeer could not have foreseen all of the fortuitous developments that would rain down upon him and Genzyme over the next decade. It is beyond doubt, however, that he recognized many of the pieces on the chessboard as they either had or would appear.

And he was yet to demonstrate his mastery of how he would leverage and maneuver them.

But for Termeer, this would prove to be his life's fulcral period, turning the arc of his career from corporate executive to entrepreneur. He had either gathered or been born with the organizational skills, the charismatic leadership presence, the strategic vision, and the drive to build a major enterprise from near nothing. His destiny lay before him.

CHAPTER FOUR

Mission: Impossible

By the end of 1983, Wall Street was coming to understand that the most interesting sector to emerge in the new economy was the biotech industry.

Jim Sherblom, recruited by Henri from Bain and Company to be Genzyme's first Chief Financial Officer (CFO)—he was employee #22—recalls that "there were only six start-ups in the Boston area, and the three that were best regarded by the Bain folks were the three that had Baxter Travenol people running them."

The three were Gabe Schmergel's Genetics Institute, Bob Carpenter's Integrated Genetics, and Henri Termeer's Genzyme. But when Henri took his first measure of exactly what he had taken on, he must have felt like anything but biotech's golden boy, Baxter connection or no Baxter connection.

Sherry Snyder was going to be a problem. The mercurial Genzyme founder/CEO had often been a source of turbulence.

One day, Ginger More remembers, she got a call from England. John Littlechild had just come back from the races where the competitors' lineup contained a surprise, "I saw a horse running. It was called 'Genzyme Gene.' Isn't that a coincidence?"

Ginger More thought to herself, "Oh, my God."

The horse was a line item, albeit hard to find, in the company's budget. It turned out Genzyme Gene "was on the payroll."

Eventually, everyone knew about Genzyme Gene. It would become the stuff of company legend, but for Henri, in 1983, it was just one more, big headache, a symptom of some wrong turns.

A couple of weeks into his new job, Sherblom was in England meeting with the company's U.K. controller to prepare for an upcoming audit.

"We're going through the items," Sherblom remembers, "and he goes, 'this one you're probably going to have to think about, because the audits are going to flag it.'

"I'm going, OK, what is it? And the controller goes, 'The Chairman was here, and he was at the New Market Races. He was trying to think about how to raise Genzyme's profile, and he saw that everyone was talking about the horses that won that day. So he arranged to buy a horse from a trainer and hire the trainer to take care of the horse, and put all the feed bills and the horse costs and the trainer costs through the marketing budget and it gets consolidated up to corporate."

Sherblom said, "Well..." and came up with a plan.

"We got the auditors to agree, since the company was still private, to pass this with a footnote until we could figure out a way to resolve this issue."

Snyder had even taken Henry Blair to meet the thoroughbred, but Blair didn't know he was in the presence of quivering, snorting company property.

"I had no idea he was doing this on Genzyme's money," Blair says. "I thought Sherry had a lot of money, but he actually was very good at using other people's money."

Eventually, Genzyme Gene's—and Sherry Snyder's—luck ran out. Snyder put the horse up for auction, where it was bought by its trainer. Henry Blair and Ginger More both recalled what happened next. "When he had to get it off the payroll, it started to win!" Ginger More remembers.

But an errant racehorse was not the only strain on Genzyme's budget that Sherry Snyder would have to answer for. Jim Sherblom's first day on the job as Genzyme CFO was a stranger-than-fiction moment of truth.

"I was supposed to start on a Monday, but I get a call from Henri who says, 'you used to live in London.'"

"I said, 'Yes, I used to live in London.'"

"Can you start in London?"

"Okay, I'll fly to London on Monday."

"No, no, you need to fly out on Sunday, overnight, so you can arrive Monday morning and pick up the Chairman of the Board, Sherry Snyder."

"I said to Henri, Okay, I'm CFO, I don't know much about this job, but I guess I'm picking up the Chairman and we're going to go and meet the bank manager."

That's when Henri dropped the other shoe, Jim Sherblom recalls. "He says, 'Yes, the banks called us, we're in default on our bank line, and you need to go there because I don't think the Chairman can handle these kinds of things.'"

Jim Sherblom flew out on the red-eye, got into his rental car and drove to pick Snyder up at his hotel. "I'm wearing my Bain outfit of dark blue suit, red tie, white shirt, all proper for a meeting with a banker. And I go to pick up Sherry and he's wearing blue jeans and a bomber jacket."

"We get to the bank and the manager is dressed to the nines. His assistant makes sure we have coffee, and we're sitting around his little table in his office eating cookies and making small talk. He wants to talk about our families, he wants to talk about his family, and I think we're building some sort of relationship. Finally, after we've done this for about 15 minutes, he sets his coffee aside, looks at Sherry, and he says, 'You're in default on your bank line, how are you going to make it correct?'

"And Sherry puts his feet firmly on the floor, leans forward and says, 'Not only can we not make it correct, but we're going to need more money, or you're going to lose it all.'"

Jim Sherblom remembers turning white. He took a breath and invited Sherry to take his coffee and cookies outside the office. Sherblom closed the door after Snyder left and offered the bank manager the names of three CEOs who could vouch for him.

"I point out that this is my first day on the job, as a matter of fact, I've only been doing this for a few hours, so how much time can he give me to make things right?

"'Based on your promise,' the manager says, 'I'll give you six weeks.'"

While Jim Sherblom cleaned up the company finances, Henri Termeer's focus was on executing what had long been his dream—a new business model for a new industry. Biotech's path could be described in a line that President Reagan liked to quote from a popular movie of the time, *Back to the Future*, "Where we're going, we don't need roads."

But before he could soar in the clouds, first Henri Termeer would have to shore up Genzyme's base.

The partners had just closed a round of financing, and "Henri thought there was money in the bank," according to Peter Wirth, "but in fact most of the money went to pay debts that had been incurred. He didn't get the financial support that he had anticipated, and I think he blamed Sherry for that lack of transparency. That sort of got their relationship off to a bad start."

Unlike the other biotech CEOs, Termeer was not comfortable with the highly risky, time-consuming, discover-and-develop model they deployed in chasing new molecules/medicines.

He told his investment banker, Peter Drake, PhD, about his plan.

"Here's what we're going to do. We're going to take the profits from our diagnostic business, and we're going to plow them in, and we're going to develop a completely new model for the development of drugs."

Henri liked to say that Genzyme was the only biotech company with a shipping dock. Peter Wirth says Henri was proud that "they made and sold products.

"Henri was very conservative in how he built his business, so he would run the business as breakeven. If he could raise money, he would do that and spend it on research. If he couldn't raise money, he'd retrench and try to live off of the revenues he was generating."

As Peter Wirth reminds us, Henri was a businessman, first and foremost.

"He was totally pragmatic. He was trained as an economist. He was all about finding opportunities and then seeing what you could make with that opportunity. And he was very much about building on the assets that he had, and the assets that he had weren't that sexy. I mean, basically, you were going around collecting placentas and extracting an enzyme. This was not biotechnology, this was protein purification."

Termeer's personal style, pragmatic instincts, and financial discipline eventually came to wear on Sherry Snyder. It was "oil and water" as Henry Blair later characterized it. The two rarely saw to eye-to-eye, and Snyder had reached the breaking point, as Termeer neared his.

As a Founder/CEO, Snyder knew his time in the leadership role would not last long. Upon his hiring, he thought Termeer would be his successor. By now, that had all changed. He convinced the board in the spring of 1985 to consider others, besides Termeer, as his successor. A search was commissioned, and a process quietly began.

The search dragged on, unsuccessful in its mission. And by the following December, it all came to a crushing head.

Termeer had originally accepted his position, in October 1983, as the company's President. The expectation associated with his hiring decision had been clear. He would become Genzyme's CEO within a year of his arrival.

It had now been more than two years, and that, of course, had not happened. Termeer's patience had been exhausted. Termeer confronted the board to conclude its search and to appoint him as Genzyme's new CEO.

The company was now engaging with bankers to go public, and it was beyond doubt that Termeer would make a stronger CEO to lead the business through the process. He also had, unquestionably, the résumé to go with the expanded role of leading the company into its future.

But questions had been raised over Termeer's judgment and his unyielding commitment to glucocerebrosidase, his risky Gaucher R&D program. Sherry Snyder could not support Termeer's decision to pursue its development, let alone his appointment as his successor.

Snyder and Drake were out for dinner at Boston's Four Seasons Hotel the night before the IPO drafting session would begin. Drake could not believe what he had heard Snyder just say, "There's gonna be a board meeting tomorrow and we're gonna fire Henri Termeer."

Drake's reaction was swift, "I was crestfallen and bewildered."

When later asked why Snyder had decided to make this recommendation, Drake remembered Snyder's answer, "He didn't think Termeer had the energy level to be a CEO."

The next day, on December 5, 1985, the board made its own decision. Sherry Snyder would be named Genzyme's Non-Executive Chairman and asked to step aside. Henri Termeer was appointed President & CEO of Genzyme. Snyder would relinquish the Chairmanship in 1988, resigning from the board altogether to pursue other interests. On his departure, Termeer would consolidate the Chairman's title and serve as Genzyme's Chairman, President & CEO for the next 23 years. The two hardly spoke ever again.

The period leading up to the crucial 1985 board meeting had been a rough couple of years, and the lack of clarity around leadership succession had only created further challenge for Termeer. But Henri now had Genzyme's reins, and, in that important respect, the tide had turned.

From that point forward, Henri Termeer put his chips squarely on glucocerebrosidase, based on the results he had seen firsthand in Roscoe Brady's lab. He would also acquire businesses to diversify the portfolio and take pressure off the earnings statement. Whenever anyone would raise a doubt, Henri Termeer would go back to Brian Berman and glucocerebrosidase, "We know it works."

Years later, in one of its storied case studies, a Harvard Business School professor described Genzyme's Gaucher development program and the meeting at which Henri Termeer decided to proceed with the development of Ceredase.

Termeer would regularly bring his scientific advisory board (SAB) together for all-day Saturday meetings—informal brainstorming sessions to thrash out the many opportunities that lay before the young company.

On a fall Saturday in 1985, the advisory board focused on whether Genzyme should invest in the research that Roscoe Brady had been leading. There were a lot of problems.

"For one whole day," the Harvard study wrote, "the scientists debated the issues with management, trying to answer three questions: Does it work? Is it safe? And could it be made profitable?

"On the first question, the scientific advisors were doubtful, arguing that there was no strong indicator that the results from Brian Berman's one case could be generalized... . The debate about safety was equally troubling... . Publicity about the risks of HIV and hepatitis C had led to growing public concern about products derived from human tissue, leading the scientists to suggest it would be more prudent to wait until biotechnology could create a recombinant version.

"Finally, there were major questions about whether a business could be created. Some raised concerns about accessing a sufficient supply of placentas, while others focused on the huge investment required to develop this product. Henry Blair, conservative by nature, worried it could bankrupt the company; Sherry Snyder also had argued against the proposal."

The advisors had been joined that morning by Ginger More of Oak Investments, Genzyme's founding venture capital partner, and Henry Blair, Genzyme's founding enzymologist. The pair sat quietly, their discomfort growing as the advisory board members focused on two potential deal breakers.

First were the significant problems with the drug therapy itself.

Here was, in their eyes, an unproven experimental product that had only gotten results in one of eight patients, hardly a ringing endorsement for its efficacy. Also, importantly, a year's supply of the product required that Genzyme somehow find 22,000 postpartum human placentas to treat just one patient.

But even so, the market was way too small—they thought maybe 5,000 patients worldwide. The whole idea was on its face preposterous.

Where on earth would Genzyme be able to collect all the placentas needed to treat the patients with the disease? And then how on earth would Genzyme be able to harvest them and create an FDA-approved product?

How would people be able to pay for it? (Early estimates put the cost per patient per year into the tens of thousands of dollars per year, far outstripping what the most expensive drug products sold for in 1984.)

For the advisory board, the problem was obvious: How would Genzyme turn all the investment and the cost of R&D, process development, and production into a viable business?

The answer was equally obvious, at least to them.

But Henri's answer, too, was obvious, just not the same one.

"We had met for a whole day and asked ourselves, 'Should we push this forward or should we move on to something else?' They decided that it was not right to go on because they thought that gene therapy was just around the corner, and, in any case, we couldn't get enough placentas. To solve the dosing problem, they said, 'You need recombinant technology, and if you get that, you may as well go right on to gene therapy, because that's the next proper step.' I didn't take their advice. I respectfully continued to push forward, based on my observations of the patient."

Or, as he said some years later, "When scientific advisory boards composed of noted people recommend not doing something, they're not always right."

Scott Furbish, PhD, a member of Brady's NIH team and one who later joined Genzyme, would say, "I would like to ask Henri how he had the guts to make that decision."

It was Termeer's bold bet, and one that would change his life and legacy forever.

Henri Termeer's answer was rooted in what he had seen with his own eyes. Brian Berman, still a patient in Genzyme's experimental trial, was proof enough that the science worked—they just did not fully understand it yet.

There were several thousand other patients out there besides Brian. "It is our obligation to find and treat them. If we don't do it, who will?," Termeer would maintain.

Roscoe Brady and his team conjectured the answer to the drug's lack of efficacy in the trial's seven nonresponders could be found in the way the drug was dosed. The other patients, mostly in their teens, were being given the same dose as Berman, a small, 30-lb child. Greater size and weight required larger doses. When that change was made, their hunch was it would show a similarly powerful response.

Within months, using a dosing regimen that adjusted for body weight, another experimental trial was begun. This time, the group responded and their health improved markedly.

That further confirmed to Termeer what he had observed in the dramatic recovery of young Berman—and, laughed one Genzyme veteran, "contaminated his thinking" for the rest of his long tenure at Genzyme.

"Once we had a dose response," Termeer recalled later, "I simply refused to consider that it might not work, as long as patients were diagnosed correctly. That's the advantage of not being burdened by knowledge. When you're an economist, you can make that assumption. A biologist would think of five million reasons why it might not work. I tend to be right by pure luck. There were a lot of people who told me this was crazy and tried to direct us differently."

The next looming question, that of product supply, was not as easily solved. The need to collect six tons—SIX TONS—of human placentas on an annual basis to serve the world's 5,000 Gaucher disease patients was a huge concern. And then there was the matter of developing a process that would consistently meet stringent FDA standards and yield cost-effective results.

If there was good news to be had, it was that Termeer and several of his leadership team members had worked or consulted for global Fortune 500 plasma fractionator, Baxter. The idea of taking human tissue—in Genzyme's case, placentas—and extracting an enzyme from the tissue was not novel. Baxter did it every day, all around the world. So why could Genzyme not find a way of doing the same, simply relying on a third party to help them?

For help, Termeer turned to one of Baxter's European competitors, Lyon-based Pasteur Merieux. He knew the Merieux organization. It was one of the early pioneers in plasmapheresis and in developing blood products from whole blood.

In the mid-1980s, Merieux was one of the largest placenta collectors and blood and plasma fractionators in the world. The company collected five and a half tons of human placentas every year, including 70% of all the live births in the United States. It also, after processing each placenta, threw away massive amounts of placental material as medical waste.

After approaching the leadership at Merieux, Termeer directed his senior production operations colleague, Dr. Geoffrey Cox, based in London, to fly over to France, where he was met by his counterpart and Merieux representative, Jacques Berger. The two of them met over several trips and hatched a partnership that ran from 1987 to 1996 to produce from human placentas the bulk material to be used in Genzyme's Gaucher enzyme therapy.

Termeer later recounted, "They put placentas through beautiful wine presses to extract the liquid, which went to a plant for fractionation. The tissue was discarded. We needed what they were throwing away. I went to Pasteur Merieux and said, 'We want the tissue that you discard. We'll give you money to build a plant.'"

The alliance was extraordinary in several ways. First, there was the simple fact that it was established and operated on the basis of a handshake. There was, for its first three years, no master agreement, no contract, no binding agreement—just good faith and an understanding that Merieux would produce the material on a cost-plus basis and receive a small royalty on Genzyme's ultimate net sales.

Jim Sherblom said, "It always seems to be one of the oddest things, why they would give us access to that enzyme on the terms they gave it to us, because they could have, basically, held us up and taken them from us."

Second, was the astonishingly fortuitous discovery that Merieux was already collecting more than one-half of the world's human placentas to derive blood products for its core business.

Finally, the $5 million in capital costs to build a Genzyme-dedicated manufacturing site was in place, provided by Genzyme's IPO in 1986.

Merieux invited Genzyme to build an annex in a small field adjacent to and connected with their main Marcy l'Etoile fractionation site outside Lyon.

Alison Taunton-Rigby, PhD, Genzyme's Senior Vice President, Biotherapeutics, remembered the site as "this big shed area."

"It was dark, dirty. Everyone was wearing Wellington boots up to their thighs. They wore plastic aprons and were covered. Because they were handling infectious blood, we didn't go into the main processing facility. We thought it was too dangerous.

"We built a shed on the back side with a small, single door, and they passed the tissue through to us. We partially processed the tissue in France, made an extract of it, and got it safe enough to ship to the States."

To finish the processing, Genzyme opened what Geoffrey Cox remembers as "a little plant" on Binney Street. Alison Taunton-Rigby remembered a small room with "presses and big vats."

The Binney Street site provided the team with its own independent production facility, enabling it to produce glucocerebrosidase, which after downstream modification and vialling, became suitable for intravenous administration to the patient. Up to that point in time, the team had relied on NIH to produce and finish the drug product.

"We set up a small pilot plant there because there was no other place to go," said Dr. Taunton-Rigby. "That's where we finished it off, developing all the protocols and procedures to actually make the enzyme." Years later, after Ceredase's approval, the small facility would be swiftly approved by the FDA. Genzyme was getting closer to achieving its mission.

In parallel with the product amendments, clinical trial successes, process developments, and the establishment of the Pasteur Merieux alliance, Termeer confronted other struggles more typical of young, aspiring, but unproven biotechnology enterprises.

The company's financial condition was, at times, precarious. Through its IPO, the company was able to access the public capital markets in June 1986, raising $28 million. But within 12 months, it was burning through its cash reserves to pay for new employees, the Marcy l'Etoile site, and clinical trial development.

For help, Termeer turned to his old friend, Joe Cohen, Chairman at Wall Street boutique, Cowen & Co., to raise additional funding. Cohen in turn looked to Jim Sherblom who had an idea. He knew that others had been successful by raising capital through real estate limited partnerships (LPs). Together they reinvented the LP's mechanics to create one for the purpose of funding Ceredase's R&D. The innovation would win Sherblom a place on the cover page of *CFO* magazine.

Throughout the summer of 1987, the company failed to make much headway in its desperate effort to finance the promising new drug's development. Finally, with a few weeks to go, Genzyme was resorting to pitching individuals and brokers in smaller cities like Albany, New York.

Genzyme flew Dr. Robin Ely, who was eight months pregnant with her fifth child, up from LaGuardia to Albany to help make the case. Two of her first four children had been diagnosed with Gaucher disease. She had already become something of a spokesman for the nascent Gaucher patient group, and she was deeply grateful to Termeer.

The Genzyme group set up a Kodak carousel slide projector to tell the story of Brian Berman, and according to his mother, "capture the hearts and hopefully then the pockets of the people who were listening."

Robin came in and spoke to the moneyed crowd about the possibility that her unborn child might be born with Gaucher. Then she grabbed Termeer's arm and said, "Henri, I think I'm in labor."

Termeer hurried her outside and told her, "I'm a derailed economist. You're a doctor. I can't help you." She then said, "It's OK. It's probably a false sign." Which in fact it was—from bouncing around in a small plane to Albany.

Meanwhile, back in the meeting room, Robin's false labor had produced an electric effect. As Jim Sherblom remembers, "We probably had a hundred people in the room, and they're all writing checks because they figure this woman who understands the science is putting herself at risk."

By the end of September, the company had its money. Enough subscriptions were sold that, according to Jim Sherblom, "we could feel like this was actually going to happen. Now, all the other Cowen offices started paying attention to the fact that one of their offices had made a big commitment here. That was the turning point."

This was a very lucky break because days later, on "Black Monday," October 17, 1987, the stock market suffered an historic collapse, losing 25% of its value in one day—still the single greatest one-day crash in the history of the Dow Jones Index.

Termeer's R&D LP deal had closed in the prior week. The funds were in the bank, and the company continued to forge ahead with its prized Gaucher program.

As Termeer would later confess, "The marketplace fell apart. Had we not been able to do that transaction ... we would have given up."

After all of the land mines that Genzyme had dodged through 1987, one would have thought that the R&D LP might have signaled a change in Henri Termeer's luck. He was proving himself a master with rabbits and hats. He would need them.

AIDS-related complex (ARC) was now clearly seen as a threat to the world's blood supply. ARC was designated a rare disease by the FDA in 1986. It is a blood-borne, viral disease, and where there is blood, there is the potential for the transmission of ARC.

What that meant for Genzyme was the introduction of a host of viral contamination issues and concerns for its budding, flagship franchise. At first, the extent of the disease's footprint was unknown, and the scope of its threat was likewise unknown. But Termeer well understood from his Baxter days that Genzyme's source of glucocerebrosidase, placental tissue, was at risk of contamination.

Regulators, Gaucher patients, and their advocates quickly determined that the imperative of treating Gaucher patients far outweighed the risks of their contracting ARC. So Ceredase would be allowed to continue to advance.

However, the pressure for Genzyme to protect its Gaucher patient base mounted rapidly as ARC spread. The answer would later be found in developing a new, second-generation, recombinant version of Ceredase (which had not yet even been FDA-approved). At that moment, Cerezyme was born, and the next chapter in Gaucher treatment opened.

CHAPTER FIVE

Delivering Hope

The atmosphere at Genzyme's Kendall Square headquarters was tense. Along with Roscoe Brady and Norman Barton, Henri Termeer and Alison Taunton-Rigby were in Rockville, Maryland, presenting to the FDA Advisory Committee (AdCom). The committee had assembled to assess the clinical trial readout and to vote. Would they recommend approval of Ceredase for the treatment of Gaucher disease?

It was October 22, 1990, and everyone at Genzyme knew this long-awaited day would be remembered for years to come—one way or the other. Even the NASDAQ stock market had taken notice as trading in Genzyme's shares had been halted until the AdCom's decision was released.

Jack Heffernan said it felt like waiting for a host at the Oscars to say, "The envelope, please."

Finally, the phone rang. Heffernan, Termeer's HR head, was in his office. Henri was calling from a pay phone in the dingy basement of the Ramada Inn, down Rockville Pike near the FDA. After three hours of deliberation in the Embassy Ballroom, the Advisory Committee had just adjourned for the day. It was 5:10 in the evening.

"I'd been waiting for the call," Jack Heffernan recalls. "My office was right next to the cafeteria. We weren't going to assemble the group if it was bad news.

"Henri calls," Jack Heffernan remembers. "I answer, 'Henri?'"

'Jack?'

'Well, Henri?'

"All I heard was he's sobbing his ass off. I'm saying, 'Henri?'

"He said, 'Yes, yes.' I said, 'Good, good.'

"I get off the phone, I go to whoever knows how to run the PA system ... 'Attention please' ... I was like, this was it!"

The two men knew what this meant. Thousands would be spared the physical devastation, wheelchair dependency, and early death that accompanied Gaucher. As was the case for Brian Berman, it meant no more enlarged spleen, bone abnormalities, severe anemia, and dangerously low platelet counts. It meant a near-normal life.

Following the Advisory Committee's recommendation, the FDA would go on to approve the drug the following April. But this had been the critical step, vanquishing the naysayers who said it could never be done.

The electricity of this moment was in the air. Between Dr. Roscoe Brady's epic toil (begun in 1966 and aided by Norman Barton, Scott Furbish, John Barranger, Elizabeth Neufeld, and others in Brady's NIH lab) and Termeer's unwavering determination, conscience, and courage, it had been decades in the making. It was a pivotal, and perhaps the seminal, moment in the crusade to follow—to discover and develop new therapies for the worldwide community of rare disease sufferers, their family members, and their caregivers—an estimated half a billion people. The door of hope was finally opening.

The agency had assigned Solomon Sobel, MD, the FDA Division Chief for Metabolic and Endocrine Drugs, as the lead reviewer for Genzyme's application. He had been the lead reviewer on the approval of the first ever genetically engineered product, Humulin (recombinant insulin). And, as the Genzyme team would later learn, he was a member and devoted congregant of the Beth Sholom Congregation, an orthodox synagogue in Potomac, Maryland, the house of worship at which the Berman family prayed. Another of the clinical trial participants, a little girl from South Africa, also attended services at the temple.

"Solomon Sobel watched those little kids for six months while they were being treated, and they physically looked more energetic," Alison Taunton-Rigby remembered. "They no longer had bleeding issues as they had before, and they didn't have bruises all over their body. He watched them get better. He was watching all the time, but we didn't know that. That's why he was willing to go to bat for the treatment protocol."

Lasting three hours, the meeting had been tedious and statistical in its subject matter, clinical, and, at times, antagonistic in its tenor. The committee had interrogated the details surrounding not only the results but the design of the pivotal study, conducted at NIH, and created and executed by Norman Barton, MD, PhD. He was Dr. Brady's trusted lieutenant.

Brady, the physician who led the NIH's Developmental and Metabolic Neurology Branch, was the lab chief and the visionary; Barton was the expert in translational medicine. Twelve patients diagnosed with Gaucher disease were enrolled. Barton oversaw their dosing, monitored their health and well-being throughout, and gathered and reported the results. The trial had commenced April 24, 1989.

Barton vividly remembered Termeer and the AdCom meeting, a milestone in the history of orphan drug therapy development, "Henri stood up and went to the microphone at the end of the meeting. It's unusual for CEOs to speak up at these meetings. They're high pressure events because of what's at stake. Most CEOs tend to leave the proceedings to the science/medicine people and the statisticians. But Henri made his pitch. This was a miraculous, beneficial drug. He said basically, 'Come on, approve this drug.' It was pretty impressive.

"He really kind of bet the company on this one."

Abbey Meyers also knew the importance of the moment. "Pharmaceutical companies were accustomed to thinking big: Big patient populations would lead to big profits. They never explored the need for treatments of uncommon diseases. Orphan drugs were facing an industry deeply entrenched in this model... . New innovative people willing to take risks needed to become involved."

Running the gauntlet of the pivotal FDA clinical trial that preceded and ultimately led to Ceredase's approval, although speedy by nearly all measures, was not without anxiety. After all, this was new territory; enzyme replacement therapy was cutting edge, novel, unproven.

The trial's success was chronicled in *The New England Journal of Medicine* on May 23, 1991. Participating patients ranged in age from 7 to 42; five were male, seven were female. The trial was conducted over a one-year period for each of the patients selected to participate. All twelve patients responded to therapy. The trial had been resoundingly successful.

As Phil Reilly, MD, described it, "After a year of therapy, the hemoglobin concentration improved in all the patients, the plasma levels of glucocerebrosidase decreased in nine patients, and the spleen shrunk in size in all... ."

So with Ceredase's FDA approval in April 1991, hope had been delivered, and with it, a cascade of developments followed. A groundswell of patient advocacy, started by Abbey Meyers and others after passage of the Orphan Drug Act in early 1983, grew exponentially. The approval also triggered industry interest in targeting the development of therapies for other rare disease targets, such as cystic fibrosis, Fabry disease, Pompe disease, MPS 1, muscular dystrophy, Tay–Sachs disease, and Niemann–Pick disease. Regulators and lawmakers stood up and took notice. The use of special financing vehicles like R&D limited partnerships to finance innovative medicines mushroomed. Venture capital caught the wave and began aggressively founding and funding innovation-driven enterprises that deployed recombinant DNA biotechnology, some of it focused on the new area of orphan drug development.

For Termeer, this era of his leadership, the 1990s, anchored his position as a founding father of the orphan drug movement and a champion for rare disease patients. It also provided him with the foundation of confidence, and to some degree invincibility, that propelled his development into the charismatic biotech CEO he became.

As David Meeker, Termeer's successor at Genzyme upon its acquisition by Sanofi, reflected fondly, "Henri wasn't born great. But he had a larger than life manifestation, completely out there. Everything was possible. He challenged conventional thinking. He didn't jump for the easy conclusion. He was an intensely curious man who was interested in life and all it offered. His early days in the Gaucher community shaped him forever. He had a natural ability to connect with people personally, especially with employees, patients, and their families. The authenticity of his warmth and interest in people was independent of rank and structure. Genzyme helped him unlock his inner self, and he led with a style to which people responded."

The accolades poured in, but Termeer was not one to bask in the glow. In his modest, self-deprecatory way, he set up a file at the office, a box labeled in large handwritten black Sharpie ink, "Big Head," into which he would deposit the puff pieces as they rolled off the printing presses. He would have none of it. It was a way for him to humor himself. Although proud, he was not a promotional or boastful man. By the time he died, there were two Big Head file boxes among the scores of other boxes in the basement of his Marblehead manse.

Termeer's colleagues enjoyed a front row seat as Termeer grew immensely during this period. They saw a man who was leading a movement with patients at its core. They also saw a leader who became the ultimate human capitalist—attracting, retaining, developing, and mentoring an army of employees who shared his vision for Genzyme and his passion for patients.

There was perhaps no better illustration of Termeer's growing impact than his service to BIO (known today as the Biotechnology Innovation Organization), the biotechnology industry's trade association that he helped form in 1993. It was the result of a merger of two related, but somewhat competitive, associations that represented the interests of biotech companies in the public square. It became obvious that it was in everyone's interest to unify the industry's voice. Termeer was its first Vice Chairman and ascended to its Chairmanship two years after its founding.

Along with Kirk Raab, BIO's founding Chairman and Genentech's CEO, Termeer played an integral role in attracting Carl Feldbaum, the incumbent Chief of Staff for Senator Arlen Specter, to join as the organization's first President. Feldbaum would go on to serve under six BIO Chairmen through 2004.

Termeer and Feldbaum met in December 1992. It was not a one-on-one visit, but rather a search committee meeting in which Feldbaum was interviewed for the job. Feldbaum commented, "Henri held back, but not much, and weighed in with more strategic, philosophic, if you will, issues... . Unlike other members of the search committee, he had had some connection with legislative affairs. He was not naive and not a newbie ... for most of these folks it was a black box."

Feldbaum was ultimately selected and began work in January 1993, the week of Bill Clinton's inauguration. To Feldbaum, he found his initial relationship with Termeer to be "not so much wrong-footed, but more a little off-footed, in a way, and challenging. It turned out to be extraordinarily creative for me."

Later in their relationship, after a dinner attended by Termeer, Feldbaum, and Feldbaum's wife, Laura, a former Marshall Fund executive, Carl recalled his wife's remarks, "'Carl, just remember, these Dutch people explore the unknown world without charts,'" attempting to explain Termeer's curiosity, his unpredictability, and his willingness to consider things in ways that others would not have dreamed.

Feldbaum and Termeer made an exceptional team. They shared deeply held values around global and domestic health policy and together they lifted each other, lobbying and cajoling major figures on Capitol Hill and in the White House to protect the interests of not only the innovators but the rare disease patients for whom they were developing therapies.

"When Henri became Chairman of BIO, we arranged a call, one hour, each week," says Carl Feldbaum. "We always had it on time, and it virtually always lasted an hour.

"The agenda was, 'Carl, you tell me what happened and what you did this week.' And then he would walk down the beach with a pair of tweezers, turning over every piece of sand."

Carl Feldbaum did not mind this level of supervision. "Embedded in what others might call micromanagement were comments and conversation about what we valued in people in terms of experience, in terms of credentials, and in terms of what we were trying to accomplish.

"I called it a séance. It wasn't a meeting. It wasn't a telephone call, it wasn't a report. No, it was a séance. You got the sense of sitting at a table, everybody's hands on the table. There would be a whole spectrum of stuff we would talk about, including some really high-level strategic thinking. He was seeing, which I was not, five or six or seven steps ahead."

Termeer was blessed with a self-confident social ease and grace. He had an uncanny ability to connect with people of all walks of life. He had an amazing ability to create a bond that many described as "being in the moment" with Henri. You had his full attention. It was precious time. He was an amazing listener. He had empathy. He cared.

One of his long-time Genzyme colleagues, Tomye Tierney, described Henri in this way, "His insight, his compassion, and his ability to make you feel like you were the only person in the room. He could read you like a book... . He was inspirational... . He could make you feel like crap, but just his spirit—he made you feel like you were walking on air."

These gifts were ones that served Termeer throughout his life, and in the context of the 1990s, they lubricated his interactions with regulators, government officials, patients, their families, and, of course, patient advocates. He was able to bring them all together.

Patient advocates, especially those within the rare disease community, were organizing and creating their own place within the U.S. healthcare ecosystem. They were a force to be reckoned with and an increasingly important voice within the U.S. health policy debate. No more vivid an illustration was their support of innovations directed at rare disease communities.

Rhonda Buyers, former CEO and Executive Director of the National Gaucher Foundation (NGF), worked with Termeer for nearly 20 years, joining the NGF in 1994. She described the patient advocate experience in the early days, "Initially, the foundation was looking to be able to help educate people all over the country. We combined forces to reach people and get them the care they needed. We worked on HIPAA laws, lifetime insurance caps, and assisting patients with reimbursement. Patient families, especially those with more than one Gaucher patient, would quickly reach their lifetime cap. We were concerned about that. And pricing was another big concern. But these were miracle drugs, and Ceredase and Cerezyme served as a blueprint for many other drugs.

"I recall when I first started traveling around the country and meeting people that had the disease. People were in wheelchairs. They were using crutches and all sorts of assist devices. And years later, they weren't. That's why I always say they are miracle drugs. I recall being at patient meetings, standing up at the podium, crying, because I was just emotional with seeing all these kids that were just running around and having a great time. They all had Gaucher disease. It was just amazing to me to see that transformation."

For Genzyme, having delivered Ceredase as a newly approved therapy for the treatment of Gaucher disease, the journey was just starting. There was much left to be done. Directly ahead, there was the imperative to develop and deliver Cerezyme.

Cerezyme was Genzyme's answer to the limited availability and growing threat of HIV contamination of human placental tissue. It was a recombinant biotech drug that was expressed from mammalian cells that had been transfected with new genetic machinery. The process eliminated the viral threat that Ceredase faced.

Genzyme's ease in shifting gears from Ceredase to Cerezyme was in no small part attributable to Termeer's familiarity with blood processing from his time at Baxter's Hyland Division and its production of blood-derived Factor VIII. He knew the risks of contamination and how to mitigate them. This familiarity with biologicals manufacturing became defining, and in some ways differentiating, for Genzyme. Through its acquisition of Integrated Genetics in 1989, practically overnight Genzyme became one of the world's leaders in the use of cutting-edge mammalian cell culture processes to make recombinant biotech products. This was a new capability that became essential to producing recombinant biological drugs in commercial scale quantities. These capabilities were crucial to the introduction of Cerezyme, approved by the FDA in 1994, as Ceredase's replacement.

Beyond the realm of Gaucher disease, Termeer had already set his sights on not only other rare diseases that fell within the same family as Gaucher, the family of lysosomal storage disorders (given the humorous acronym of LSDs), but also a wide swath of healthcare products. Although three of Genzyme's LSD targets, Pompe disease, Fabry disease, and MPS 1, were clearly "druggable," their product launches were not expected until well into the next decade.

This led Genzyme to a strategic crossroads. Years before, in the mid-1980s, Termeer had sketched out his vision for Genzyme. It was to be "quite deliberately organized as a diversified business." It would raise $500 million, operate through four global business units, and fully integrate into R&D, manufacturing, and marketing/selling. Indeed, Genzyme was not, at least initially, designed to be a rare disease enterprise, and the determination of Genzyme's eventual identity as the global leader in rare diseases required help from a competitor's challenge that served to crystallize the company's strategic direction.

In the late 1990s, a little known upstart biotech concern, Transkaryotic Therapies, or TKT, recognized the rare disease arena as potentially a very attractive segment in which to build a company. It had seen Genzyme's success with Gaucher, and TKT was developing a new therapy for Fabry disease. At the time, Genzyme was not investing aggressively in the LSD therapies that would comprise its future rare disease portfolio. Although Ceredase and Cerezyme had hit the market, even Genzyme did not yet fully appreciate the economic potential of the ultrarare disease business proposition.

TKT announced in June 1999 that it had filed for orphan drug designation of its new agent, later branded Replagal, as an enzyme replacement therapy for the treatment of Fabry disease. This proved to be the shot across Genzyme's bow that would demand a swift competitive and strategic response.

The two firms were developing a nearly identical drug, and the winner of an FDA approval would receive orphan drug exclusivity for seven years, blocking the other from the U.S. market. In very short order, Fabrazyme would become a top institutional priority for Genzyme.

The companies battled, filing lawsuits against one another claiming patent violations. TKT and Genzyme both raced to obtain marketing approval, and each got EU approval for their respective products on August 3, 2001. In the United States, both companies filed registration dossiers with the FDA. Courts issued opinions, and the divide deepened as the stakes grew.

Rich Moscicki, Termeer's Chief Medical Officer, remembered those times well, "It was quite a battle. Henri had been thinking, early on, that Fabry disease afflicted such a small number of patients—was there room for two companies? He approached Richard Selden, MD, PhD, TKT's President & CEO, to explore the possibility of the two firms working together and sharing the market. And Selden said this to Henri, 'No, I'm going to tell you how it's going to go, Henri. You're going to leave Fabry to me, and I won't come after you on Gaucher.'"

Dr. Moscicki also remembered, "Henri left the meeting, went to me, and said, 'Win.'"

Now bitter rivals, Termeer's fierce competitive instincts were revealed. He would joust with Richard Selden until the tussle was decided.

The short, but rather intense hostilities abated on February 11, 2003 when it was announced that Dr. Selden would step down from his post. The FDA advisory panel had recommended 15-0 against TKT's Replagal product in a mid-January meeting, opting instead to recommend Genzyme's Fabrazyme. TKT's shares fell more than 25% on the day's news. Genzyme basked in the glow as its next rare disease product launch beckoned.

On April 24, 2003, it was officially announced that Genzyme had received FDA approval for Fabrazyme, and with it, the promise of U.S. market exclusivity.

The episode, while only four years in the making, was an important one as it would give definition to Genzyme's strategic priorities, underlining the centrality of rare disease therapies as the company's signature strategic anchor. It also positioned Termeer as the rare disease industry's most central figure.

In 2003, with the imminent U.S. launches of Fabrazyme (2003), Aldurazyme (2003), and Myozyme (2006), Genzyme was on the march. Its net sales had already crossed the $1 billion mark. The firm was diversifying its revenue sources and expanding around the globe; it had entered more than 50 markets. Total employees numbered around 6,000.

With all that was going right, however, there continued a raging debate, begun in the early to mid-1990s, that severely challenged Termeer, Genzyme, and the orphan drug industry in ways that it had not anticipated a decade earlier. The debate? It was drug pricing.

CHAPTER SIX

Duty, Honor, Patients

How did rare disease patients become central to the life of a man raised in a family of shoemakers? The two do not usually line up. In Henri Termeer's case, the answer could be traced back to the Bonn Hemophilia Center where the young Baxter executive had walked the halls in the late 1970s, visiting with patients, talking with their families, and meeting their physicians.

He had seen the desperation of the young German boys as they faced living each day, reliant upon a drug therapy. For these rare disease patients, it was the difference between life and death. And the anxiety for parents and loved ones was equally gripping.

Like Brian Berman, whom he would encounter several years later, Termeer had seen the lifeline with his own eyes. And from this period forward, there was no choice. When confronted in the early 1980s with the numerous obstacles that biopharmaceutical companies faced in developing treatments for rare disease patients, his mantra was and remained, "We must do this." It became his own personal crusade.

Within months of joining Genzyme, his obsession with helping patients became core to the company's culture. Some would describe his passion as "maniacal." Others would choose "humanitarian" or "caring." Still others would say "courageous" or "unwise" as he pursued products for which there appeared to be little commercial potential.

But when aggregated, these points of view reverted to the mean. And as Genzyme's plot unfolded, Henri Termeer earned recognition as the supreme industry champion for rare disease patients. In a pharmaceutical world that was chasing blockbuster, billion-dollar rainbows, he and those whom he led would choose to help the forgotten orphaned patients for whom there was no treatment and often no hope.

As Genzyme's development progressed, the FDA's orphan drug program would continue to break new ground. The biopharmaceutical industry's pipeline of new therapies continued to swell with new product candidates.

To be sure, in Genzyme's infancy, the pathway forward was far from easy. Termeer and Genzyme were facing challenges and opportunities that had never been faced before. And these were not ones found in a strategic plan sitting on some shelf in a dark closet, collecting dust. They were fresh, abundant scientific and medical advances that were being developed every day in search of cures for patients with unmet medical needs.

From the mid-1980s through the early 2000s, Big Pharma scorned rare diseases. The idea of a specialty field sales force of nine people was, at the least, unconventional. For some it was a foreign concept. Big Pharma was building several hundred person armies to market and sell to the mass market orally administered remedies for medical conditions such as gastric ulcers, high cholesterol, heart failures, and allergies or asthma. Few thought Genzyme's model had much commercial potential. They had concluded it was not worth the trouble and the rewards too small.

As Peter Drake, the Kidder Peabody biotech analyst, described Termeer, "I remember Henri in those days. You have to realize, at the time, a strategy built around rare diseases was considered anathema by the big pharma industry. They thought it was laughable ... that you could actually develop a drug that treated not a lot of people, and be able to price it at a premium and build a huge organization on it."

In the late 1980s, Genzyme was approaching the end of the long march to gain FDA approval and launch Ceredase, which would become the company's anchor asset. The dose-ranging study of the experimental product was nearly completed, and its therapeutic efficacy had more or less been established.

Termeer took intense interest in the study and was known to have a microscopic focus on the twelve patients being treated. He not only got to know their names but many of the details surrounding each particular case. It was at this time, as Ceredase was being prepared for launch, that he demonstrated that he was not only patient-centric, but data-centric.

Termeer would astonish his colleagues with an extraordinary recall of patient detail. It was uncanny. Hemoglobin, check. Platelet count, check. Liver volume, check. Height Z-score, check. Termeer would know the pathology of every case. It would drive his GMs crazy. After all, he was the CEO, not the product manager. How could he keep track of all that minutiae?

The field sales representatives, case managers, and others who accumulated such data often provided him with the scoop. After all, Termeer would be prone, at the drop of a hat, to dip down three or four layers into the Genzyme organization in search of answers to almost anything. The org chart meant next to nothing when he was on a quest for answers.

Although their formal title was Clinical Science Associates, the sales representatives' capabilities and buy-in would be critical to Genzyme's and Ceredase's success in the field. Mind you, these were not traditional pharmaceutical pill peddlers. These were the elite. Many had been hired away from Genentech and other leading biotech firms. Known internally as "CSAs," they would operate on the front line, identifying and bonding with patients and their physicians.

They were trained by Scott Furbish, who had recently been brought on board to Genzyme from NIH. Roscoe Brady would school the team in therapeutic nuances of the drug's administration. Norman Barton would also play an important role in their education. He had led the pivotal clinical trial that had provided the foundation for the drug's FDA approval. Collectively, they heightened the CSA's understanding of the drug's clinical performance.

Lance Webb, one of the first CSAs hired at Genzyme, recalled those precious months when Ceredase was being launched, "In those early days, when you connected with these patients and they found out that you understood Gaucher disease, because nobody else did, you could be on the phone for three and four hours, just answering questions and telling them this and that, because it was like they latched onto you for information they could never get. I spent many, many a night going to the dinner table with my phone in my hand, eating dinner, and still sitting there a couple hours later. The dishes were clean and everything's gone and Barb's essentially eaten by herself. But that was what it was like."

These good-natured first responders of sorts would ensure that Gaucher therapy was delivered to those who needed it. The CSA group, initially nine in number, was comprised of individuals who had many years of experience in the pharmaceutical/biotech industry. These men and women thought outside of the traditional ways of presenting new, innovative therapies to physicians, other healthcare professionals, and patients. Their personalities brimmed with pragmatic optimism and a "can-do" vibe.

In those early days, no other biotech or pharma company organized or deployed its sales force as did Genzyme. The CSAs got deep in the needs of their patients, deeper than their counterparts at other pharmaceutical and biotechnology companies. Their interactions with the patients were high touch and offered a heavy component of education around the disease state itself.

Lance Webb remembers his first meeting at the home office with Henri Termeer. The CSAs had been brought in for a few days of briefings, discussions on setting up treatment centers, going over logistics, marketing plans, and other aspects of the business.

"Henri came in to talk to us. Of course, based on my experience, I fully expected to hear the speech about, 'Okay, we've got to make a sales target of X to get us through this quarter or that quarter, and these are our projections for the year,' but no, Henri said, 'The best thing you can do for us to succeed is to be the patient's lifeline.' I thought, 'Wow, that's interesting.'"

Lance Webb saw the uniqueness of Termeer's approach, "Henri had a humanistic side that you don't find in most CEOs, especially today. He was thinking about the patients. He understood that this is a different business. This is a one-on-one, patient-oriented business. He wasn't just focused on numbers. He was focused on patients. I think a lot of his humanism came from his mother."

Rhonda Buyers knew each of Genzyme's CSAs and appreciated their dedication, "They were incredible, generous people. And they knew everything about the patients—medical history, financial circumstance, you name it. There was enormous trust between them and the patients and their families. They deeply cared about their patients. It was the environment that Henri Termeer created."

Supporting the CSAs were Genzyme's case managers. These caring individuals would do battle with the insurers and others who would be asked to reimburse for Ceredase's administration. They would also help needy patients find the means to pay for the normal day-to-day costs that would accompany the drug's administration—things like infusion costs, travel expenses, babysitting, and hotel accommodations. When the money ran out and patient's basic needs—think electric bill or grocery tab—were not able to be met, Genzyme and the National Gaucher Foundation (NGF) would help them find support. These patients, in times of need, would also seek support from NGF's Care Plus Program, which would provide individuals with assistance in meeting their insurance premium payments. Years later, restrictive laws would be enacted, citing potential or real conflicts of interest. They would forbid many of these practices. But in those early days of rare disease patient care, this was the way it was done. Common sense, humanitarianism, and respect for human life trumped other considerations.

Lance Webb found the case managers to be key, "Without the case management department to advocate on behalf of the patient with these insurance companies and clerks, a lot of patients would never have been approved. The clerks were advocating for the insurance company, and they did not understand the disease, and in some cases, their initial response was to deny coverage for the therapy. The case managers were almost always successful in gaining approval from the insurance companies after submitting documentation of the patient's symptoms and explaining the therapy."

Termeer's imprint on this approach was palpable. Buyers put it, "It all started at the top. He was like a proud papa, always very kind and caring. And he was a pioneer. I can't even imagine having done what he did. He stuck to his guns. He was a force to be reckoned with, and when he said something, people listened. We all kind of knew he was one of those special people."

Kathleen Coolidge, a case manager who had joined Genzyme in 1997, described the gestalt, "He treated Genzyme's patients like VIPs, and his leadership team mirrored him. He set the tone and they followed. I had never experienced the level of respect and appreciation and desire to know someone's story like Henri had with patients. It was a value that he had. To explain Genzyme and its products, he would tell patient stories, and it was not artificial. He was genuinely moved and people could feel that. The way he talked about the business and the way he talked about the disease was like talking about an individual person. There was something innate in him that enabled him to communicate that patient's individual story and bring Genzyme to a personal level."

Genzyme grew dramatically in the 20 years after Ceredase was approved, and many initiatives sprung up to connect the now-sprawling corporation with its patients. One such initiative was Genzyme's Expression of Hope art show, a global disease awareness program that featured works of art created by rare disease patients. Started in 2006, it was Termeer's way of thanking the patients and their families while also recognizing that their lives were defined by more than having a rare disorder. The program also enabled the Genzyme team to better appreciate the journeys of the patients they were working to serve.

These events bonded the various rare disease patients and their families to each other, bonded patients to Genzyme, bonded employees to their mission, and bonded Termeer to all who partook in them.

In communicating their importance, Genzyme's head of patient advocacy, Jamie Ring, offered this account, "We would invite patients from all over the world to participate in helping raise rare disease awareness by creating and submitting their original artwork. Over time the program received hundreds of submissions from dozens of countries. The program included art ranging from finger paints done by young children to photographs and still life paintings from seasoned adult artists affected by a lysosomal storage disorder."

Ring explained the goal of reaching as many employees as possible, "We later took the exhibit over to the Allston Landing site where the bioreactors are located. We also created note cards and reprints of the exhibit pieces, and employees were able to have access to these. You would find the images throughout the different facilities on people's walls and desks to serve as a reminder of why we were at work every day. The program was profoundly motivating for employees and such a classic representation of Henri's genuine care and concern for the global patient communities we treated."

Termeer sought to foster similar connections in Genzyme's international offices. He would travel the world with Sandy Smith, his head of International, visiting patients in far-flung corners of the globe. They connected with families throughout Europe, in Latin America, and across Asia from Japan to Singapore.

One venue that hosted several patient events over the years was the company's manufacturing site in Geel, Belgium. Sandra Poole had played a leadership role in the site's development, leading its operation from 2004 to 2009, enabling her to work closely with Termeer and observe his leadership.

In October 2004, shortly after Poole had assumed her leadership duties, Termeer was making the rounds visiting the company's production facilities in Waterford, Ireland; Haverhill, United Kingdom; Marcy l'Etoile, France; and Geel. When visiting Geel, he would often bring members of the Genzyme executive team and always find time to break away, drive up to Tilburg, pick up his mother, and bring her down to meet his colleagues and, of course, patients. She was one of the "VSOPs," very special old people, as Henri called them.

Mary Termeer took an intense interest in Henri and the rousing success that was Genzyme. She and Henri would talk weekly about the company and its fortunes. Like her son, she also took deep interest in the well-being of Genzyme's patients.

Poole got to know Mary Termeer through these visits. Mary was the grande dame, the belle of the ball. Poole reflected on Mary's impact on these events, "We would involve her in the community relations and present to her what we were planning. It was funny because here we were... . Henri's mom is sitting in the middle of the table. She was the focal point, and he was devoted to her... . It was like she was in charge."

At one of the Geel events, Poole remembered that Mary had met Maryze Schoneveld van der Linde, a courageous Pompe patient in her thirties who had battled the neuromuscular degenerative effects of the disease. It was another of the lysosomal storage disorders for which NIH's Roscoe Brady, as well as other prominent clinical researchers, had been searching for a cure. Genzyme was racing, with Brady's help, to develop a new therapy for her rare genetic disorder. Mary leaned over to Henri and could be heard whispering, "You have got to find a way to get her that medicine."

A year later, indeed, Genzyme received FDA approval for and launched Myozyme for the treatment of Pompe disease. The approval capped a multiyear effort to find the best among several competing approaches, made more noteworthy by the fact that in the early years the Genzyme team had believed Pompe disease, which was much rarer and more complex than Gaucher, might be beyond even their capabilities.

Schoneveld van der Linde had been diagnosed in January 1978 at the age of eight. It was a day she would remember vividly. Following an examination of her symptoms and case workup, her physician, a pediatric neurologist at a leading Dutch teaching hospital, explained to her, "You have a serious disease. It's very rare, and you actually are the first child that I've ever diagnosed with this disease. Unfortunately, we don't have a treatment." Her life would never be the same.

At the time, Schoneveld van der Linde could walk, ski, skate, and ride her bike like other children. But as the disease progressed, her neuromuscular capabilities suffered. Her mom would try to encourage her, "Come on Maryze, we are soldiers, and we have to do this together."

But the progressive disorder was unforgiving in its cruelty and she, in a low moment, several years later, would sit in her bed, crying, asking, "Why do I need to live this life?" Her mother would respond, "Well, Maryze, I don't know, but I think you were born to do something with your life and to be special and to make a difference in this world."

In the 1980s, this was the world of pediatric medicine for the rare disease patient community and their families. Schoneveld van der Linde would eventually need the assistance of a caregiver to live her life, but her unyielding spirit continued to drive her even as her physical capacity diminished.

After earning a Master's degree at the University of Leiden in cultural anthropology, she would go on to form a patient advocacy organization, Patient Centered Solutions, which would become a voice for Pompe and other rare disease patients worldwide. She also had founded the International Pompe Association (IPA). Her motto, "the impossible is possible," signifies her optimism in the face of her physically devastating condition.

She commented, "You learn that collaboration is key in this whole process. And not only collaboration with patients, but also collaborations with scientists, the medical society, but also the authorities, because I was talking to parliament members and ministers. I told them how important it is to look not only at Pompe, but all rare diseases."

Shortly before Myozyme's approval, while she was attending an IPA convention in Boston, Termeer invited Schoneveld van der Linde for a visit at his office. Her mother joined them. Henri could see that Schoneveld van der Linde's condition was deteriorating rapidly, and he said to her, "Maryze, I don't want to lose you. Because you are important and we need people like you to make this happen."

These heartfelt scenes of patient support and interaction were repeated in countries around the world. Belgium, China, Brazil, Egypt—these were places where Termeer took personal interest in patients, visited them frequently, and partnered with them and their families to improve their condition.

One touching visit came in India, where Termeer visited a family that had moved far from home so that their daughter, Nidhi, could get Genzyme's therapy for Pompe. Her parents explained that she had started therapy too late to walk unaided, but that they were so grateful she would live that they had begun an effort to get others diagnosed earlier. Perched on the girl's cot in their cramped quarters, Termeer realized she was the same age as his own daughter, and to her delight took down her address so they could be pen pals.

Gatherings and trips, however, were not all that Termeer did for the patient community. He challenged the Genzyme organization to deliver more, especially in the realm of connectivity, helping with things like diagnostics, registries, and networks. In the pre–social media years of the company's development, it was a challenge to even know who confronted the realities of living with a rare disease, much less to connect with them.

Termeer frequently met with John King, a long-time product manager, to quiz him on his work. King would play a substantial role in the rollout of five new products, two of which were the LSD products launched in the spring of 2003, Fabrazyme and Aldurazyme.

"Henri was always asking patients how we could serve them better. And he would listen. If I was in a meeting with Henri, I had better remember what was said and get it done. The responsibility rested on my shoulders. And Henri would always circle back within a week and ask what I had done about it," said King. "Where are we with this?" Termeer would ask. And King better have an answer when he did.

Termeer's contributions to the rare disease patient community were widely recognized by leaders outside Genzyme as well. One was Dr. Marlene Haffner, an internist, hematologist, and Rear Admiral in the U.S. Public Health Service who additionally served for 20 years in the FDA with oversight for its orphan drug products program. Haffner, a tireless advocate for rare disease patients, managed the orphan drug designation process, "Orphandom," as she coined it.

Her recollections of Termeer were informed and illuminating, "Henri was quiet, physically unimposing, understated, dynamic, brilliant; he cared. He made the commitment that any patient with Gaucher's who could be treated with his drug would have access to it, regardless of whether it could be paid for. I don't know of anyone else who has done that. It was a unique philosophy."

Richard Pops, Chairman and CEO of Alkermes, a Boston-based neuroscience concern he has led since 1991, and earlier a biotech banker at Paine Webber, had a front-row seat in observing Termeer's 30+-year impact on shaping an industry, "I met Henri in the late 1980s. Back then, biotech was a small cottage industry. There were very few principal players. Henri was one of those.

"The contrast between Genentech, Amgen, and Genzyme was quite striking. If you went to Amgen or Genentech, there were scientists and laboratories and an ethos driven by an excitement around the power of recombinant DNA technology. When you went to Genzyme, it was Henri in this warren of offices in an office building in Chinatown. It didn't feel like science at all. It felt like science was a tool to build a business. Henri was quite promiscuous in terms of the different types of technologies, and business structures, and science that he would accumulate. The Baxter guys were the business guys. Henri was clearly the most transactional and business focused of that generation of CEOs.

"Before Termeer, the whole ambiance of the pharmaceutical industry was so different. Nobody would have contemplated a $300,000 medicine, because nobody would develop a drug for a small patient population. And nobody would conceive of pricing something at those levels. But the era of the blockbuster changed it all. Henri was the first one to figure out how to solve for being a blockbuster, irrespective of the number of patients. His was a completely novel way of thinking about the whole thing. The narrative shifted from a pricing model based on cost of goods to value to the patient."

It was this shift in the fundamentals of the rare disease marketplace that gave rise to a whole host of new issues in the patient community. Not only were patients now seeing previously untreatable diseases treated, they were witnessing a related, logarithmic shift in drug pricing.

Abbey Meyers, the NORD cofounder, was arguably Termeer's equal in her passion for the discovery and development of therapies for rare disease patients. She appreciated his passion for the treatment of rare disease patients, but recoiled at the prices charged for them.

Meyers offered these reflections on her relationship with Termeer and his leadership of a movement he created with her, "I was always impressed at how he was revered by his employees. They admired him so much. No matter what, through thick and through thin, Henri was there. We had this healthy respect for each other even though we disagreed with each other. He was always very friendly. Except he was immovable."

In a mid-1990s visit to Genzyme, Meyers recalled an encounter with Termeer, "I met him as I was leaving the building. I stopped and said, 'You know, I really have to thank you, because there were so many people who said they'll never manufacture an orphan drug because you're never gonna make enough money from it. You took a chance. I hate your pricing, but you proved to the rest of the industry that you can make money, you can make a profit on an orphan drug that's only sold to four or five thousand people... .'

"It was painful to patients, but he proved something that got other companies to turn their head and say, 'Maybe Henri was right. Maybe we should look at these drugs.'"

When asked if Henri was the father of the Orphan Drug industry, the executive who created its business model, Meyers would reply, "That's exactly right. And as a matter of fact, the last time I saw him face to face, that's exactly what I said to him."

The issue of high drug prices—especially in the rare disease segment —would explode in the coming years.

Every ounce of Termeer's persuasive capability, charisma, and self-assurance would be called upon to defend not only Genzyme's but the industry's pricing practices and policies.

On a bustling market street in the center of Tilburg, Netherlands stood Heuvelstraat 39, the home of Jacques and Mary Termeer. Flanked by the Dreesmann Department Store to its right *and the iconic Gimbrère fashion boutique on its* left, *this three-story brick building housed the family shoe shop, The Termeer Shoe Co., which occupied its first floor. The Termeer family lived on its two upper floors. Henri Termeer was born on its second floor on February 28, 1946 and lived here until 1953 when the family moved to a leafy suburb near Tilburg University on the outskirts of town.* (Courtesy, the Henri A. Termeer Family.)

Henri Termeer as a young boy was, like most, mischievous, curious, and fun-loving. (Courtesy, the Henri A. Termeer Family.)

As Henri Termeer entered adolescence, his interests turned to roller-skating, biking, and exploring abandoned German pillboxes on the sandy beaches along the Dutch coast near Rotterdam. (Courtesy, the Henri A. Termeer Family.)

From the age of 12 and for the next three years, Termeer (shown far left*) was a fanatical chess player, an avocation he pursued until his grades suffered and his mother intervened. It was an interest, however, for the rest of his life. Many who knew Termeer well attributed at least a portion of his strategic, visionary capabilities to chess. In group discussions, they claimed he was often moves ahead of everyone else in the room. (Courtesy, the Henri A. Termeer Family.)*

*Entering at the age of 19, Termeer (*left*) fulfilled his mandatory Dutch military obligation by serving for two years as a logistics officer in the Royal Netherlands Air Force, based first in Breda and later at a base near The Hague. A 2nd Lieutenant, he claimed it was an important experience that gave him confidence in his leadership abilities.* (Courtesy, the Henri A. Termeer Family.)

As one of five international MBA students in his class of 105, Henri Termeer enrolled in the fall of 1971 and graduated, at the age of 27, from the University of Virginia's Darden School of Business in the spring of 1973. (Courtesy, the Henri A. Termeer Family.)

Bill Graham, or "Mr. Graham" as he was known to his colleagues, was Termeer's first mentor at Baxter and, as its CEO, a man who had a profound impact not only on Termeer's rapid professional development and leadership philosophies but on how he later led Genzyme. (Photo taken in 1989 at a black tie dinner honoring Bill Graham.) (Courtesy, the Henri A. Termeer Family.)

*The Termeer family was close-knit. Fourth in the birth order of six children, Henri was raised by a mother and father whom he later described as being "talented in dealing with kids, in giving them disciplined input while being warm at the same time." (*Bottom row*) Jacques Termeer, Mary Termeer; (*middle row*) Ineke Termeer, Marlies Termeer; (*top row*) Roel Termeer, Henri Termeer, Paul Termeer, Bert Termeer. (Courtesy, the Henri A. Termeer Family.)*

In October 1983, as Henri Termeer joined Genzyme, he would report to work at the company's 75 Kneeland Street headquarters located on the 15th floor in the center of Boston's "Combat Zone," the city's garment district, overrun in that era by drugs and prostitution. (Courtesy, the Henri A. Termeer Family.)

On many Saturday mornings throughout the mid to late 1980s, Genzyme's leadership would meet with BioInformation Associates (BIA), a group of leading MIT and Harvard faculty members (denoted in ***bold*** *type) who offered Genzyme a broad array of advisory expertise in the life sciences. (Back row) Henri Termeer, Sheridan Snyder,* ***Charles Cooney****,* ***Graham Walker****, Henry Blair, Betsy Robinson,* ***William Roush****,* ***Chokyun Rya****; (front row) Hamish Hale,* ***George Whitesides****,* ***Chris Walsh****, Tibby Posillico, Jim Sherblom; (missing)* ***Harvey Lodish****,* ***Anthony Sinskey****. (Courtesy, the Henri A. Termeer Family.)*

Charlie Cooney, PhD, a MIT global bioprocessing expert (seated left*), was instrumental in helping Genzyme in solving the production scale-up of its high-molecular-weight biopharmaceuticals, especially Ceredase and Cerezyme, the first two of Genzyme's rare disease products. Here, Cooney is seated next to Genzyme cofounder, Henry Blair. Henri Termeer stands. Betsy Robinson (seated at the* far right*), a Genzyme scientist, joins the discussion. (Courtesy, the Henri A. Termeer Family.)*

After joining Genzyme's Board of Directors in 1983 to represent BIA's 10% ownership interest, Charlie Cooney served as a Genzyme Non-Executive Director until 2011, a period of 28 years. With Sherry Snyder, Ginger More, and John Littlechild, he was instrumental in recruiting Termeer to Genzyme in the fall of 1983. (Courtesy, the Henri A. Termeer Family.)

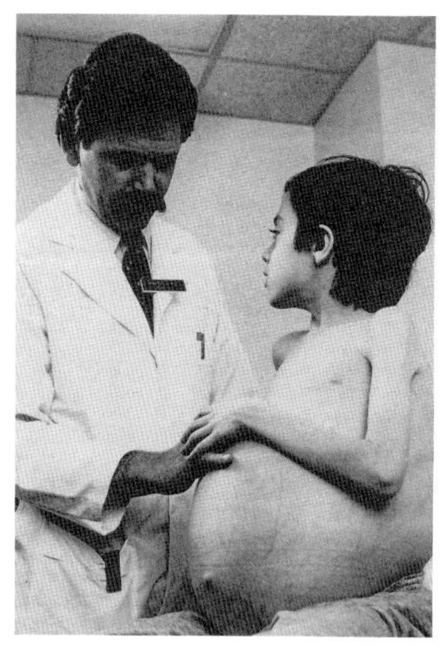

Dr. Norman Barton, a translational medicine expert, is seen examining a young boy suffering from an advanced case of Gaucher disease. These patients, if not treated, would often suffer from an enlarged liver and spleen, anemia, low platelet count, wheelchair dependency, and early death. (Courtesy, Dr. Norman Barton.)

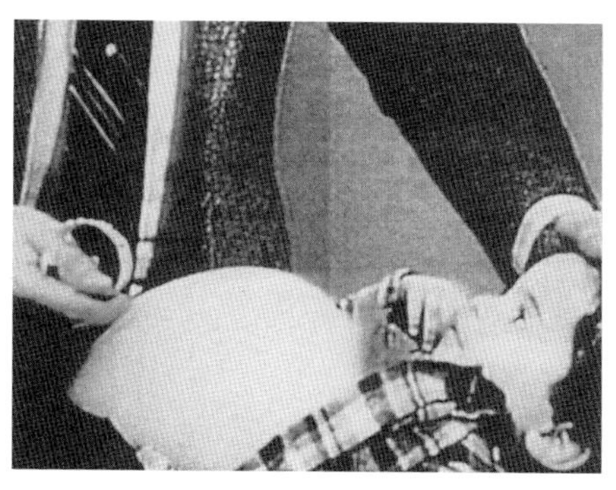

Born in late 1979, Genzyme's first patient, Brian Berman, is shown in a symptomatic stage of Gaucher disease, prior to his treatment with the company's experimental enzyme replacement therapy (ERT), which was first administered in December 1983. Dr. Roscoe Brady and other members of his lab at NIH (notably Dr. Barton, Dr. John Barranger, and Dr. Scott Furbish) presided over Berman's care.

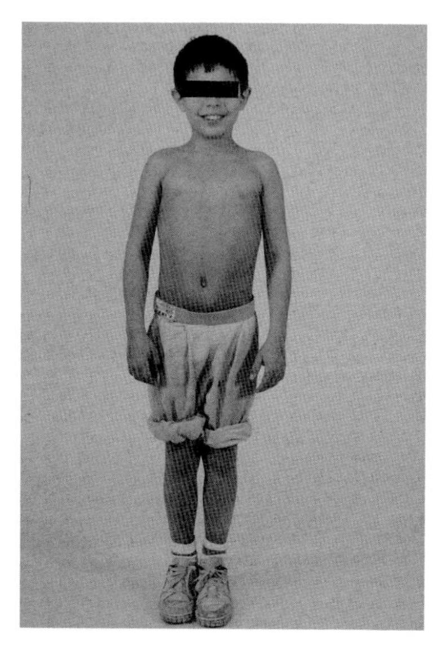

Having responded to ERT, Brian Berman is seen several years later, around the age of 7, in fine physical condition. (Courtesy, Dr. Norman Barton.)

Brian Berman, who remains on ERT, now leads a normal, healthy life and serves as the President and CEO of the National Gaucher Foundation, a foundation founded by his parents to attend to the needs of the Gaucher community.

Pasteur Merieux was instrumental in aiding Genzyme in the development of Ceredase by providing the company with access to 70% of the world's placentas from which Genzyme could extract the active pharmaceutical ingredient (API) used in producing the lifesaving drug. Without Pasteur Merieux's support, there would likely have never been Genzyme as we know it today. An estimated 22,000 placentas were required annually to treat each Gaucher patient on ERT. (With permission from Sanofi.)

Henri Termeer, Peter Wirth (then Genzyme's outside counsel at Boston's Palmer & Dodge who later joined Genzyme in 1996), and Henry Blair (Genzyme cofounder) are all smiles in the 1980s as Genzyme is developing into a global, vertically integrated biopharmaceutical enterprise. (With permission from Sanofi Genzyme.)

Young executive Henri Termeer, around 1985, the year in which he added the CEO role to his responsibilities. *(Courtesy, the Henri A. Termeer Family.)*

Henry Blair looks on as CFO Jim Sherblom and Henri Termeer hold a check representing the proceeds from Genzyme's June 6, 1986 initial public offering (IPO) of its common stock, which raised $28.3 million in gross proceeds, netting Genzyme $21.5 million after deducting deal expenses and proceeds paid to selling shareholders. Including the money raised in the offering, the IPO valued Genzyme at ~$75 million. (With permission from Sanofi Genzyme.)

Henri Termeer stands in his office, the tombstone of his IPO hanging on the wall behind. Until Ceredase was launched in 1991, Genzyme was primarily dependent on cash flows from its collection of disparate businesses and funds raised through R&D limited partnerships. (Courtesy, the Henri A. Termeer Family.)

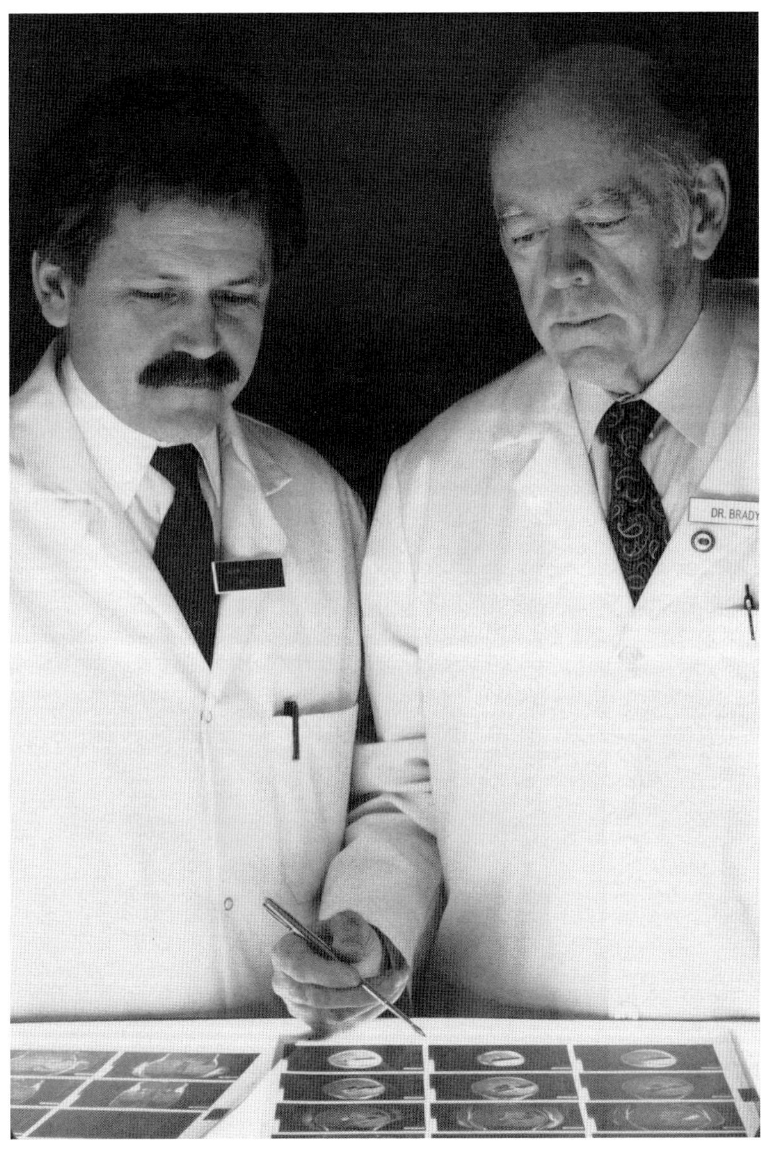

In the early spring of 1990, Dr. Roscoe Brady (right) *and Dr. Norman Barton* (left) *look over the abdominal MRI scans of twelve patients from the registration clinical trial being conducted at NIH and sponsored by Genzyme. The successful trial led to Ceredase's approval by the FDA on April 5, 1991 for the treatment of Gaucher disease.* (Courtesy, Dr. Norman Barton.)

Scott Furbish, PhD, Henri Termeer, Henry Blair, and Roscoe Brady, MD, gather in 2001 to commemorate the 10th anniversary of Ceredase's introduction, a landmark in rare disease clinical history. Although Brady led the NIH group that inspired and developed Ceredase, Furbish was the scientist who played a critical, lead role in developing the process that would modify its sugar side chains, thereby unleashing its clinical efficacy. Furbish would depart the NIH to join Genzyme in the mid-1980s to help scale the production process. (With permission from Sanofi Genzyme; Lance Webb, photographer.)

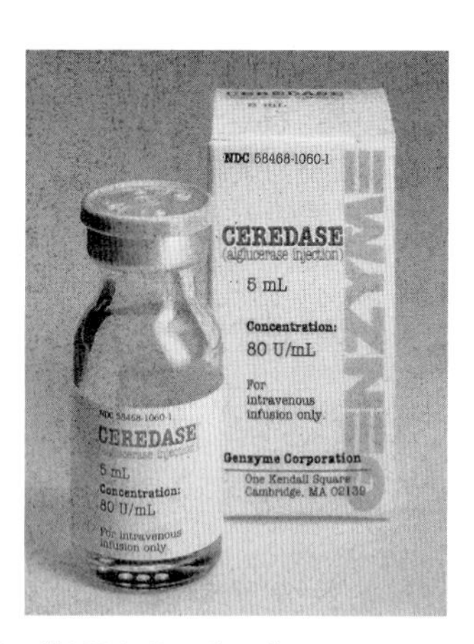

In April 1991, Ceredase became among the first biotech therapies ever brought to market to expressly serve an ultrarare condition, Gaucher disease, which afflicts approximately 5,000 patients worldwide. Derived from human placentas, Ceredase was the anchor asset around which Termer built Genzyme's global rare disease franchise, ushering in waves of biotech orphan drug therapies that followed in succeeding years. (With permission from Sanofi Genzyme.)

A Professor at Harvard Medical School and a clinical immunologist at Massachusetts General Hospital, Rich Moscicki, MD, joined Genzyme in early 1992 as its first Chief Medical Officer. Attracted by Termeer's vision of the future, his passion, and his charisma, Dr. Moscicki had a profound impact on Genzyme's future, becoming one of Termeer's most trusted advisors. (With permission from Sanofi Genzyme.)

In 1995, Henri Termeer sits between the Governor of Massachusetts, William Weld, and Genzyme operations executive, Dr. Geoffrey Cox, at the Grand Opening of its Allston Landing production facility on the Charles River. The facility was built for the express purpose of making Cerezyme, the recombinant version and successor product to Ceredase. (Courtesy, the Henri A. Termeer Family.)

Termeer, speaking at the Allston Landing Grand Opening, knew from his Baxter experience that Genzyme needed to create a new recombinant DNA production process and a new manufacturing site to provide Gaucher patients with a therapy that avoided the risk of infection from blood-borne viruses. (Courtesy, the Henri A. Termeer Family.)

The Allston Landing production facility was known around Boston as the "Cathedral on the Charles." Some thought Termeer had designed it after a European church. In 2009, because of the plant's viral contamination due to the usage in its bioreactors of infected fetal calf serum imported from New Zealand, it became the cause of "the crisis." The plant's compromised condition resulted in its shutdown, its decontamination, and temporary supply outages of Genzyme's most critical, single-source products, including Cerezyme and Fabrazyme. Corporate earnings were pressured, investor activists took notice, and the crisis precipitated a difficult chapter in Genzyme's and, by extension, Termeer's history. (Nick Wheeler, Photographer.)

*By 2000, Termeer had become one of biotech's leading spokesmen, routinely accessing the halls of power in Washington and often accompanied by his aide-de-camp, Lisa Raines, who met a tragic death on 9/11, a passenger on American Airlines Flight #77. (*Top*) Termeer pictured with President William J. Clinton; (*middle*) Termeer pictured with Speaker of the House Newt Gingrich and Lisa Raines (*right*); (*bottom*) Termeer pictured with President George W. Bush while receiving in 2005 the National Medal of Technology and Innovation on behalf of Genzyme for the company's "pioneering dramatic improvements in the health of thousands of patients with rare diseases and harnessing the promise of biotechnology to develop innovative new therapies." (With permission from Sanofi Genzyme.)*

Henri Termeer and Dr. David Kessler, FDA Commissioner (1990–1997), converse after meeting in Boston. (Courtesy, the Henri A. Termeer Family.)

Termeer and his strongest legislative ally, Senator Edward M. Kennedy, the "Lion of the Senate," discussed health reforms often, and not only including those pertaining to rare diseases but also to those that related to FDA reform, drug pricing, and biosimilars. They also enjoyed the water and could be seen sailing from time to time off Hyannis Port on Kennedy's 50-foot wooden Concordia schooner, Mya. (Courtesy, the Henri A. Termeer Family.)

Termeer became well known to Wall Street as Genzyme's fame and fortune grew throughout the 1990s and 2000s. Genzyme's half-decade flirtation with NASDAQ-listed tracking stocks, abandoned in 2003, only served to heighten investors' awareness of the company. (Courtesy, the Henri A. Termeer Family.)

Genzyme was named to the Fortune 500 in May 2010, a leadership achievement of which Termeer was justifiably proud and which he memorialized by framing the magazine's cover page and hanging it in his office at the Yellow House. (Courtesy, the Henri A. Termeer Family; photo by John Hawkins.)

*During its heyday, the Laguna Niguel meeting, hosted annually in Southern California by Kleiner Perkins venture capitalist Brook Byers (*far left*) and investor/advisor Steve Burrill (*far right*), was among the better attended biotech CEO gatherings. Pictured here in 1979 with Byers and Burrill stand four titans of biotechnology—George Rathmann, founder and CEO of Amgen (*second from left*); Ernie Mario, former CEO of Alza Corp. (*third from left*); Fred Frank, legendary Lehman Brothers Vice Chairman and healthcare banker (*fourth from left*); and Henri Termeer, Chairman & CEO of Genzyme (*second from right*). (Courtesy, the Henri A. Termeer Family.)*

Henri Termeer and his head of International, Sandy Smith (left), would travel the world, seeing as many as 15 Country General Managers in any given week. Here they enjoy some downtime together at the Singapore Zoo. (Courtesy, the Henri A. Termeer Family.)

For 20 years, Termeer drew support from Tomye Tierney (left) who architected and led Genzyme's compassionate use programs in many developing nations around the world. Here, pictured in Singapore, Tierney and Termeer enjoy the company of two chimpanzees. (Courtesy, the Henri A. Termeer Family.)

On October 7–8, 2002 in Bethesda, Maryland, Genzyme sponsored a gala event to honor Dr. Roscoe Brady for his lifetime of scientific and medical achievements. Convening attendees from the world over, "Brady Fest," as this two-day tribute became known, was capped by Termeer's presentation to Brady of a Steuben glass eagle at its climactic, black tie dinner. A clinical research legend, especially in lysosomal storage disorders, Dr. Brady passed away at the age of 92 on June 13, 2016. (Courtesy, Lance Webb.)

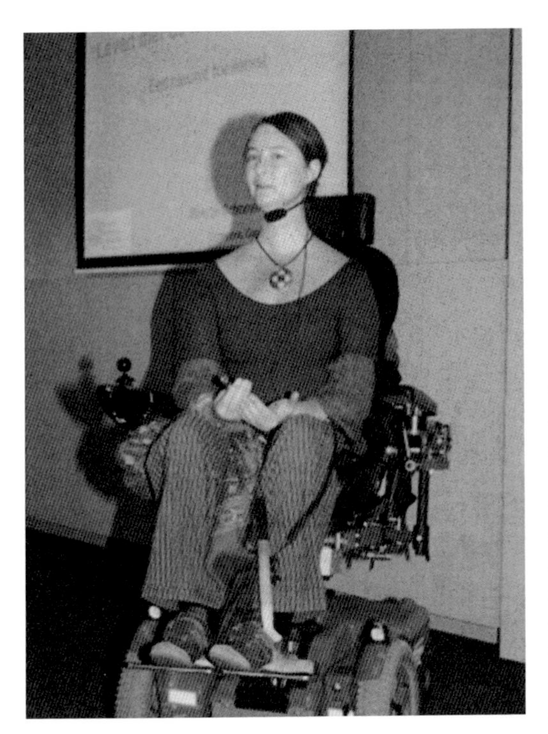

Maryze Schoneveld van der Linde, a Dutch-born, wheelchair-dependent Pompe disease patient diagnosed in 1978 at the age of 8, is now Director of her own patient advocacy organization, Patient Centered Solutions. Marveled by her courage, Termeer and she would meet periodically to discuss ways therapies could be developed to improve the lives of Pompe patients worldwide. (With permission from Sanofi Genzyme.)

Termeer would travel the globe to meet rare disease patients, delivering a message of hope and encouragement. Here he can be seen, squatting to eye level, as he visits with a young patient. (Courtesy, the Henri A. Termeer Family.)

Henri Termeer would often visit not only Genzyme's patients in far-flung places, but his company's production sites as well. Here he can be seen with Sandra Poole, who oversaw Genzyme's production site in Geel, Belgium from 2004 to 2009. Also pictured (middle) is Henri's mother, Mary Termeer. Poole would humorously describe Mary as being the center of these visits' attention, to the point where it was almost like "she was in charge." (*With permission from Sanofi Genzyme.*)

Mary Termeer shakes hands with Her Royal Highness, Astrid, Princess of Belgium in 2004 at the Grand Opening of Genzyme's substantially updated and enlarged Geel site. (*Courtesy, the Henri A. Termeer Family.*)

Jim Greenwood, Biotechnology Innovation Organization (BIO) CEO, is seen presenting Termeer in 2008 with its Biotechnology Heritage Award, given annually in conjunction with the Science History Institute at BIO's annual convention to a biotech leader who has helped "to heal, fuel and feed the world." It is among the most coveted and prestigious of the industry's awards for meritorious service. (With permission from James Greenwood.)

When asked to describe his old friend, Henri Termeer, whom he had met in 2005, Deval Patrick, former Governor of Massachusetts, described him as a "Citizen in Full," and a man who could see his neighbors' dreams as well as his own. (With permission from Sanofi Genzyme.)

On January 24, 2011, in a private dining room at the Waldhotel in Davos, Switzerland—over dinner and an exquisite bottle of Pomerol, 2001 Château Pétrus—Chris Viehbacher and Henri Termeer came to terms in merging Genzyme into Sanofi. Later as a gift (pictured here), Viehbacher gave Termeer a second bottle of the same wine, inscribed on the left side margin of the bottle's label, "Best wishes, Chris, March 2011." It resides in the Termeer wine cellar to this day. (Courtesy, the Henri A. Termeer Family.)

*On the morning of February 16, 2011, Termeer and Viehbacher (*left*) convened a joint press conference in Boston to announce the merger of Sanofi and Genzyme, a transaction that had taken nine months to consummate and had been agreed by the Boards of both companies. To acquire Genzyme, Sanofi offered $20.1 billion in cash plus an additional $3.0 billion in contingent value rights. They heralded the deal as a "new beginning." (With permission from the* Boston Herald.*)*

Termeer (seated) *takes it all in as his successor, David Meeker, had assumed the mantle as Genzyme's next Chief Executive Officer, upon the closing of the Sanofi acquisition, April 8, 2011.* (With permission from Sanofi Genzyme.)

Termeer mentored scores of his young leaders, and as they left Genzyme to run their own businesses, many adopted credos that drew from values that had been core to Genzyme's. One such typical credo statement, found today in the lobby of Cambridge biotech company Blueprint Medicines and led by a Termeer mentee, Jeff Albers, espouses five themes embraced within the Genzyme culture. (Provided by and courtesy of Blueprint Medicines.)

After retiring from Genzyme in 2011, Termeer set up an office in the Yellow House, which had been built in 1890 at the tip of Marblehead on the harbor. A wonderful old house with a warmth and charm all its own, he and his wife, Belinda, had purchased it several years prior. (Scott Booth Photography.)

On October 2, 2012, the Termeer family joined Massachusetts General Hospital leaders, faculty, and staff to celebrate the opening of the Henri and Belinda Termeer Center for Targeted Therapies. Made possible through a generous initial donation of $10 million from the Termeers, this first-in-human clinical trials center for novel cancer therapies has played a global leadership role in the development of several new therapies that have changed the standard of care for patients around the world-a testament to the vision and commitment of both Henri and Belinda Termeer to making a meaningful difference in the lives of patients. (Pictured left to right*) Peter L. Slavin, MD, President, Massachusetts General Hospital; Daniel A. Haber, MD, PhD, Director, Massachusetts General Hospital Cancer Center; John Murphy, patient, Massachusetts General Hospital Cancer Center; Adriana Termeer; Belinda Termeer; Henri Termeer; Keith T. Flaherty, MD, Director of Clinical Research, Massachusetts General Hospital Cancer Center. (Courtesy, Dr. Daniel A. Haber.)*

*Termeer took great pleasure in helping others start their own initiatives. So it was with Harvard Medical School's Biomedical Science Program, led by Dr. Joan Reede (*left*), which began as "Joan's Fantasy." It grew into one of the nation's most successful and respected programs in identifying, supporting, and mentoring minorities who chose to enter health sciences careers. (Provided by Joan Reede; photo by Jeff Thiebauth.)*

Henri Termeer served for eleven years on the MIT Corporation's Board of Members, the Institute's governing body. For six of those years, he served on the Board's Executive Committee. Here, he and Board Chairman Robert Millard discuss the day's events. (Photo by John Gillooly.)

Having just received an honorary degree from the University of Twente on November 25, 2011, Henri Termeer is pictured conversing with Her Majesty Queen Beatrix of the Netherlands at the celebratory reception that followed the conferral ceremony. (Courtesy, the Henri A. Termeer Family.)

Having joined the board as its Deputy Chairman in 2007, Termeer was elevated to Chairman of the Board of the Federal Reserve Bank of Boston (FRB Boston) on January 1, 2010, a position he held through the end of 2011. In these roles, Termeer led the Boston Fed through the global economic crisis as his corporate life was being upended by Genzyme's crisis at Allston Landing. Here Termeer is pictured with other members of the FRB Boston board and across from the institution's CEO, Eric Rosengren, who stands on the far right. *(Courtesy of Eric Rosengren.)*

Henri and Belinda Termeer were avid supporters of the Boston Ballet. They are pictured with Mikko Nissinen (left), its Artistic Director, at the Ballet's 2010 Spring Ball, held that year at The Castle, an historic, repurposed armory. Belinda Termeer served on the Board of Trustees of the Boston Ballet from 2002 to 2014. She nurtured Henri's interest, which intensified once their daughter, Adriana, took up dance. Henri Termeer joined Belinda on the Board in 2012, overlapping her service for nearly two years and serving until his passing in 2017. (Photo by Eric Antoniou.)

Henri Termeer and Belinda Herrera were married on August 1, 1998, after a long, multiyear courtship. Pictured here, they celebrated their nuptials in the living room of their oceanside home in Biddeford Pool, Maine. The ceremony was joyous and intimate. (Courtesy, the Henri A. Termeer Family.)

Many would chuckle when learning about Henri and Belinda's honeymoon, which was a camping trip in the White Mountains, accompanied by Henri's 12-year-old son, Nicholas. They spent much of it swatting mosquitoes and grilling hot dogs. (Courtesy, the Henri A. Termeer Family.)

Henri and Belinda had one daughter, Adriana, who was the apple of Henri's eye. Many described her May 20, 2017 eulogy of her father as being among the most poignant they'd ever heard. (Courtesy, the Henri A. Termeer Family.)

Henri, Belinda, Nicholas, and Adriana are photographed in Biddeford Pool, Maine with Woody, the family Labrador retriever. (Courtesy, the Henri A. Termeer Family.)

Although spread out across several countries, the Termeers had a long tradition of large, regular family reunions in the Netherlands. (Courtesy, the Henri A. Termeer Family.)

Mr. and Mrs. Henri A. Termeer were married nearly 19 years before his sudden, unexpected passing on May 12, 2017. His life was celebrated at a memorial service held on May 20 at MIT's Kresge Auditorium in Cambridge, Massachusetts and a day later at a funeral mass at Our Lady Star of the Sea Church, Marblehead, Massachusetts. His remains are interred at Waterside Cemetery in Marblehead. (Courtesy, the Henri A. Termeer Family.)

Sailing off into the sunset, Henri Termeer loved the sea and his Hinckley sloop, Tijuba. *(Courtesy, the Henri A. Termeer Family.)*

CHAPTER SEVEN

Into the Lion's Den

Leading up to Ceredase's launch in April 1991, Termeer and his marketing team were faced with the ordinarily delicious prospect of introducing a new, breakthrough therapy that possessed miraculous benefit for patients. Ceredase was one of the kinds of products that most field sales organizations loved to sell, an innovative new therapeutic to treat an unmet medical need.

But there was an unnerving reality that went along with this one. Ceredase also carried a price tag to match its high expectations, something that the marketplace had never encountered to anywhere near this degree.

The notion of a product that was estimated to cost around $250,000–$300,000 per patient per year, depending on their weight, was simply unheard of.

Hospital administrators (and their boards) always worried about money. They did not want to take ownership of an $18,000 vial only to have it later unreimbursed by the payers. Because after it left Genzyme's shipping dock, it was theirs. And if the patient was on Medicare, as many Gaucher patients were, a large chunk, about one-half, of the therapy's hospital bill would be delayed in payment because of a lengthy government-imposed reconciliation process.

And there was one other catch. Some physicians did not see enough evidence. Although the 12 patients—100% of those on treatment—in the NIH registration trial had shown therapeutic benefit from the drug, it was hardly the depth of evidence that would normally accompany an FDA approval. Statistically speaking, the case was thin.

As Termeer looked at the marketplace and the "show me" attitude that confronted him that spring, he could see the monumental challenge he faced; launching a highly effective but super-expensive product with scarce clinical data in a tiny market in which only 30 patients had yet been identified.

Like most challenges Termeer would encounter, especially those involving patients, he accepted this one willingly and dug in. By now, he had proven he was hardly a leader to shrink in front of long odds or stiff opposition. He undertook the battle that lay ahead because he deeply believed he was right. It was society's responsibility to treat these patients, whatever the cost. And if he did not do it, who would?

One partial answer could be found in the word "alignment." When Henri Termeer deployed foundational cornerstones in how he built Genzyme, two could be found in the mission alignment and cultural alignment of the organization he assembled. Not only his executive team, but the hundreds below them, hire by hire, had been built around acceptance of, and commitment to, the company's mission. And the candidate's cultural fit was as important in the hiring process as any other, maybe more so. The team aligned at two of the deepest levels of any business—mission and culture.

So naturally, as a result, among the first items on Termeer's tactical list for the launch plan was related to human capital—the hiring of a sales force, his team of CSAs, to go to market. To begin, he would divide the U.S. market into nine territories, hire great people from top companies, and populate his ranks with seasoned veterans who culturally fit Genzyme. This first generation of CSAs would prove not to be shy in accepting difficult assignments. And its members had the imagination, the passion, the courage, and the sensitivity to relate not only to physicians and hospital leaders, but also to the long-neglected rare disease patients and their families who would, presumably, go on Genzyme's new enzyme replacement therapy for life.

Initially, in the early years and mostly on the basis of price, the drug was rejected by scores of hospitals and attending physicians. As a result, it was not unusual for CSAs to suffer restricted access to physicians who could write new scripts for the drug. And when CSAs were granted an audience with hospital leaders, in scattered instances, it was not uncommon that the conversations were inhospitable and the meetings shortened.

But gradually, as patient data were accrued and the drug's outstanding efficacy profile was revealed in full, the patient and hospital communities relented.

Still, the sting of the medicine's high price lingered, and accusations of price gouging and excessive profits were being levied against Genzyme.

Working with Duke Collier, a partner in the Washington, D.C. office of the Hogan & Hartson law firm, Genzyme and Henri Termeer had received experienced advice in pricing Ceredase.

Collier had just threaded the needle in assisting Amgen's launch of its new biotech renal therapy, Epogen, a product that had been introduced in 1990 (and became known, in short, as EPO). Although considerably less expensive than Ceredase's ultimate price, EPO carried a high price that attracted enormous criticism upon its market entry. It was the second biotech drug ever launched. Moreover, Amgen had licensed away its rights to the drug except for its use in renal dialysis in the United States. As a result, the company's only "customer" was going to be the federal government, which paid for the dialysis of virtually all end-stage renal disease patients.

Collier recollected those early days in biotech's history. He had negotiated with the Health Care Financing Administration (HCFA), which was tasked with deciding the price the federal government would pay for reimbursement of drugs under Medicare.

Collier remembered his reasoning with a senior official of HCFA as they sought to set the price on EPO, "You have to set the price at $X per unit ... in order to get a return on the investment which is appropriate to the investor, which recognizes the risk of investing in biotechnology... . Let me ask you something. You're working hard, you're trying to provide for your family, you'll get a pension from the government, you have some savings. How much of your savings would you have given to George Rathmann (Amgen's Founder-CEO) for his business plan in 1980? None, right? How about a few years later when he needed more money to take his product into clinical trials? Still no money from you, right? Your money may be in T-bills, no risk whatsoever.

"Here's the thing, we have to make them (the investors) feel compensated for having taken the risk. Otherwise, they won't invest in biotechnology again. You're the only customer for now for this highly visible product in this highly visible new field. With an adequate price, biotechnology investment will be encouraged. With an inadequate price, investors will avoid biotechnology."

The official understood. The patients wanted the drug and the government wanted to have a price that would allow the company to offer it. But a return had to be offered to the investors who had taken the risk to develop the drug. Investment in a young and risky, but potentially important new industry needed to be encouraged.

With EPO's launch, U.S. Representative Pete Stark (D-CA), House Ways & Means Committee Chairman, nearly instantaneously held hearings critical of the price, but eventually his dissent and that of others dissipated as the economic reality of developing and commercializing EPO set in.

As Collier described it, "This was 'like a pure play in industrial policy.' Either the federal government rewards the risks taken by the innovator to some reasonable degree, or it doesn't.

"EPO caused a Congressional hearing with a price of six to eight thousand dollars per year. So my colleagues and I were pretty amused when we heard someone wanted to charge $250,000 per year. But I was asked to meet the CEO (Henri Termeer), who made a very persuasive case as to why the drug (Ceredase) was both necessary for Gaucher patients and necessarily expensive. Unlike most drugs, the data indicated the drug was going to be lifesaving: If we give it to them, it works. If we don't, they'll die. At the time, no one knew how many patients there would be worldwide, perhaps as few as 1,500, perhaps as many as 5,000.

"We took the same approach as we had done with EPO: What price would provide an appropriate risk-adjusted return for the time and money spent to do their research, build a plant, make the drug? We came up with $250,000, got it to patients all over the world, many of whom would not be able to pay for it.

"We built the models, did the discounted cash flow and IRR [internal rate of return] analyses. We did the same analysis as EPO. It turned out that $250,000 was actually a pretty reasonable number. Oh, and they said they'd give it to everybody around the world whether they could pay for it or not. That helped."

Collier had played a key role in the deliberations. He and Termeer became close friends. Seven years later, he joined Genzyme and its Executive Committee, staying 13 years until the company's sale to Sanofi in 2011.

Unfortunately, concerns over Ceredase's cost were not limited to HCFA, physicians, and hospital leaders. The next hurdle was the drug's dosing regimen. Roscoe Brady's rivals in academic medicine would trigger a firestorm over how often and in what increments it should be administered to patients.

One physician-scientist, Dr. Ernest Beutler, a strong-minded, confident hematologist and biomedical expert at La Jolla's Scripps Clinic and Research Foundation, challenged Brady and Genzyme. Beutler published an article in 1991 in the highly regarded *New England Journal of Medicine* (NEJM) claiming that the recommended Ceredase dosage, according to his analysis, yielded an annual cost of $382,200 to treat the average adult patient. For infants, he projected $191,000 a year.

A year later, also in the NEJM, Dr. Beutler and seven Scripps colleagues would offer what they thought was a more cost-effective way forward, a so-called "low-dose solution." Their well-composed arguments posited that the Genzyme regimen could be improved by delivering one-fourth to one-eighth as much drug, several times a week. It would be a lot cheaper, if it worked.

And in the same NEJM edition, another well-respected academic figure piled on. Alan Garber, MD, a Stanford health economist, authored a scholarly piece, entitled "No Price Too High?" that lent support to the Scripps regimen, "The higher dose (Genzyme's) was clearly very effective, but it seems unlikely that the response justifies the additional cost." (After years of further study, the low-dose regimen would ultimately be proven inferior, and the controversy would sunset.)

Two months earlier, Garber had also coauthored a report at the behest of the U.S. Office of Technology Assessment (OTA). In the report he cited the introduction of Ceredase as a case study to illustrate the challenges Americans would face in devising policies to control healthcare costs. Dr. Garber's report also cited a number of uncertainties that still existed at the time around the clinical efficacy, efficiency, and costs associated with Ceredase as a therapeutic agent for Gaucher patients. The fact that the OTA had examined Genzyme's books and determined the drug's price to be fair was not enough to stem the tide of criticism.

The matter was highly visible within the medical community and served for months as a hot topic of convention plenary sessions, academic papers, and journal articles. Patient testimonials also attested to the drug's efficacy but were not yet enough to tip the scales.

Genzyme knew there would be fire and fury over Ceredase's pricing, but it was the company's belief that the uproar would be a short-lived disruption as the drug's efficacy became known and its results proven.

Meanwhile, the field sales force's job was to find patients and get them treated. It was not to worry about reimbursement. Others would handle the money. But the question of money was always at or near the top of the objections pile. How were patients going to pay for the new drug? And how could Termeer convince hospitals to take it into their pharmacy's inventory?

At the time, Abbey Meyers was leading NORD. She too remembered well the patient community's reaction when Ceredase was launched, "Henri was absolutely a leader who was admired and held up as an exemplar of good leadership. If you spoke to anybody at Genzyme, they would follow him into hell. They absolutely admired him.

"But Genzyme did some things that got a lot of people very angry, especially the patient community. I first got involved in it because patients were calling up, screaming about the price of Ceredase. We could do a lot of things about a lot of things, but not about prices, and that's the issue that we had to deal with the most.

"So the drug gets on the market, and they had to have a price for it. And they come out with the price... . Every patient has to buy two new houses every year for the rest of their life. Because it was like $200,000 to $400,000 depending on the weight of the patient... . I couldn't believe what they did... ."

Neither could *The New York Times*. In a 1992 story on orphan drug research incentives, it declared Ceredase as "The World's Most Expensive Drug."

One furious patient had phoned Meyers, "Oh my god, I'll never be able to pay for this. This is horrendous."

Meyers, as did so many, felt helpless. After all, there was little she could do. The government had never set the price on just about anything, especially drugs. Her only recourse was to make a lot of noise, apply pressure, and see if she could persuade Termeer and Genzyme to lower the price. She enjoyed a lot of support in her efforts—from the media, payers, hospitals, other patient advocates, and from patients and their families.

Termeer searched deeply for a solution, and as he did, he would develop a first-of-its-kind plan built on his vision for rare disease patients. He knew the answer had to lie in the fact that every patient, worldwide, should have access to either private or public insurance that would cover the cost of a drug like Ceredase. To secure their access, he would need to create an infrastructure within Genzyme that would ensure that its patients received their coverage, overcoming all the obstacles payers and government agencies were expected to generate.

The model also had to be scalable and portable for application in serving the company's future needs in treating patients with other diseases besides Gaucher. It placed patients at the center of the entire organization. No drug company had ever done this before. And as it gathered traction and demonstrated its effectiveness, it would become a template for the orphan drug company of the future.

To lead this effort, Termeer would turn to an unsung hero, the man of the hour, Bill Aliski, who would be recruited to the company to address the challenge of reimbursement of what was at the time the most expensive drug in the world. He would become Termeer's first Director of Reimbursement.

For the most part, Aliski flew below the radar, but his tasks and tactics were crucial to Genzyme's success in launching Ceredase and its innovative new infrastructure model. Aliski had an unusual background for the role. He joined the company after a mid-career stop in his late thirties at Harvard's Kennedy School of Government, had served in the Peace Corps, and had been the Director of a neighborhood health center for six years in a medically underserved area near New Bedford, Massachusetts. He was part of a team delivering primary care medicine to ordinary, everyday folks in a city that had chronic and systemic medical access problems. Through his work, he became familiar with how healthcare financing worked, a body of knowledge that would be highly useful in his new role at Genzyme.

Aliski commented on his joining Genzyme, "The reason they brought me in was because they had this drug that had been recently approved and the challenge was that it had an annual cost of approximately $250,000 per adult. A lot of physicians said, 'I don't have any patients who can afford to pay that. This drug isn't going anywhere.'

"Helping patients with reimbursement became a very important function for us. We recruited case workers, nurses, and social workers who interacted with the patients by phone. These were people who knew something about insurance and medical services. They were capable of holding patients' hands and helping them navigate their insurance. This was unique. No one else had a program like this."

Aliski's team also worked closely with physicians. "We took all the obstacles to treatment out of their hands. We would do all the paperwork. We would engage with the insurance company. We would navigate the process to get the patient on therapy. We would do battle with the policy makers at the insurance companies."

The reimbursement process drew on a broad community. "Together with the National Gaucher Foundation, we developed a program to provide support and copay assistance to address patients' insurance concerns. It was a unique and extremely effective program because it helped maintain coverage for people who were at risk."

But as Genzyme was dealing with pricing resistance from the physicians, hospital administration, insurance, and patient communities, an even bigger firestorm was brewing. The full weight of both U.S. federal legislative bodies, the U.S. House of Representatives and the Senate, was about to be brought to bear.

U.S. Senator Howard Metzenbaum was among the first on Capitol Hill to draw blood. Metzenbaum, a consumer protectionist from Ohio who was fondly known in some circles as "Senator No," had taken an interest in alleged drugmaker abuses occurring with respect to the Orphan Drug Act, an act he had supported more than a decade earlier.

Described posthumously in 2008 by the *Cleveland Plain Dealer* as "a curmudgeon, the last angry liberal," he was concerned drug companies were profiteering, even possibly price "gouging." They might need reining in, he thought. He was considering the imposition of orphan drug sales caps and a trigger mechanism that would rescind the ODA's benefits to drugmakers who were abusing the law.

Metzenbaum concluded a public hearing was needed on the matter. Scores of pharmaceutical and biotech CEOs were invited to testify. Avoiding the poorly disguised trap, most declined the Senator's invitation to appear that morning of January 21, 1992. Termeer, however, was one of the two who did. Although he did not seek the spotlight, neither would he abrogate his responsibility to defend his policies and his vision for the future.

At 9:42 a.m. Senator Metzenbaum strode into Room SR-385 of the Senate Office Building to open the hearing. In his opening statement, Metzenbaum took few prisoners, "Today, this subcommittee is going to hear some dramatic testimony about these lifesaving and hope-giving orphan drugs.

It will make you stop and think about the thousands of families in this country who are being forced to pay exorbitant prices if they want to relieve the suffering of their children or loved ones. Unfortunately, too often they find those prices simply out of their reach.

"Each of us gathered in this room must put ourselves in their position. We would pay anything, including giving up our homes, to help our loved ones get the drugs they need. The prices being charged for some orphan drugs strike me as something akin to ransom. I can think of nothing so cruel as a picture of a parent who must watch a child suffer from a disease for which a medicine is available for sale, but practically unavailable."

Two hours later, facing the parabolic dais of lawmakers settled in their cushioned, dark brown leather chairs, it was Termeer's turn. He sat at the long mahogany table, a nameplate and microphone placed in front of him.

Senator Metzenbaum introduced him. Allocated five minutes to make his prepared remarks, Termeer spoke, "The question that has to be driving much of the considerations here has to be gouging; the question has to be abuse. Do we have a law that encourages abuse? The question is not (the Orphan Drug Act's) efficacy. In fact (the law's) efficacy is what we must protect here. I am interested to protect that.

"The orphan drug law can be fine-tuned, and gouging must be outlawed and we must be tough against that. I would like to work with anybody on this. We, as a company, would like to work with everybody on this, and I think we can be successful. But, please, do not put a cap because it does not deal with the issues at hand here."

As was his custom, Termeer had spoken truth to power. Metzenbaum's cap idea faded and, eventually, disappeared.

Later that year, in November 1992, Genzyme and the whole biotech industry would face their most existential threat of all. William Jefferson Clinton would surprise George H.W. Bush, winning election as the nation's 42nd President. The rapid shift in the political winds was stunning.

By the following March, two months after his inauguration, Clinton would deliver on a theme that had been central to his election: He planned to propose sweeping legislation to overhaul the nation's healthcare system. Led by First Lady Hillary Rodham Clinton, a task force would be appointed, 30 working groups convened, countless Congressional hearings held, and voices heard from every corner of healthcare.

Nearly instantaneously, biotechnology companies far and wide recognized the threat. As storm clouds darkened, Termeer would engage—full bore and without apology—in advocating the interests of Genzyme and the nation's biotechnology industry.

On November 20, 1993, through a bill sponsored by Senate leader George Mitchell and his House counterpart, Dick Gephardt, Clinton would introduce the Health Security Act, a piece of legislation that would, if passed, transform the nation's healthcare system. Among the elements proposed in the landmark legislation were price controls on pharmaceuticals, including breakthrough drugs like Ceredase. Industry leaders, biotech trade rags, and various other publications decried the proposal.

Since the turn of the year, the stock market had gradually taken into account the jeopardy overhanging pharmaceutical and biotechnology innovators. The ASE Biotech Stock Index had fallen 33%, IPOs for biotech equities had dropped precipitously, and individual share listings had taken a beating, including Genzyme's.

Indeed, one of Termeer's first acts in preparing for the battle over Hillary Care and other legislative challenges had been the hiring of Lisa Raines, who commenced her employment with Genzyme in May 1993 as its VP Government Relations.

By his side, Raines, 34, would join Termeer from BIO to serve as his aide-de-camp in Washington. She was one of the biotech industry's very first government affairs experts. From his days at Baxter Germany, Termeer had remembered and understood the role of lawmakers in shaping not only the nation's laws but the media. He would not be outflanked, and for eight of the most critical years in the evolution of biotechnology, Raines would serve him and the industry with distinction. As a pair, they would play an outsized role for nearly a decade in shaping virtually all legislation that affected the development of the biotechnology industry. (Raines would meet a tragic death on 9/11 as a passenger aboard American Airlines Flight 77, hijacked on that fateful morning by five Wahhabi terrorists and flown into the Pentagon. Termeer would eulogize her at her funeral.)

After being skewered by Senator Metzenbaum and having in a year's time become a grizzled veteran in handling contentious audiences, Termeer was angling to get more involved in the healthcare debate that was raging in Congress. As one of BIO's Vice Chairmen, he sought to testify at an upcoming hearing that was to be chaired in mid-November 1993 by Senator David Pryor. The Special Committee on Aging of the U.S. Senate had asked the question in titling the hearing, "Pharmaceutical Marketplace Reform: Is Competition the Right Prescription?"

In announcing the hearing, Senator Pryor had expressed that he was not convinced that particular breakthrough drugs, especially those covered under the ODA and including Ceredase, were reasonably priced. Senator Alan Simpson politely admonished his friend from Arkansas for not, as a matter of fairness, inviting biotech leaders, including Henri Termeer, to defend their policies and practices. Pryor claimed the speaker roster was already too full.

Termeer made his presence felt just the same. On November 12, four days before the subcommittee met, Termeer would pen a note to Senator Pryor expressing his regret for not having been invited to testify. In his letter, he would set forth ten bulleted points of rebuttal, each defending the pricing of breakthrough drugs and Genzyme's practices.

On November 22, Senator Pryor would return the favor, writing to Termeer and thanking him for his testimony *in absentia*. He would offer to include his remarks as a part of the permanent hearing record.

The record would also include an op-ed, written under Termeer's byline, entitled "The Cost of Miracles." It had been published by *The Wall Street Journal* on November 16, 1993, the day of the hearing. This exceptional piece of historical prose took dead aim at President Clinton's proposals on breakthrough drugs.

The White House knew that the healthcare industry would unleash an unprecedented attack on its Health Security Act, but President Clinton's response was too late and challenged from the beginning. The bill's scope and complexity were simply too much. A year later, the legislation would die without so much as being put to a vote.

Termeer had successfully run the gauntlet. Along with a handful of other biotech company CEOs, notably Bob Swanson and Kirk Raab of Genentech and George Rathmann at Amgen, who too were developing high-priced biotech therapies, Termeer took up the mantle and would emerge as one of the leading spokesmen of his era in defending his industry's pricing practices.

A final small storm around pricing was triggered by the absence of a price reduction when Genzyme's second-generation Gaucher product, Cerezyme, was released in the spring of 1994. One of the primary justifications for the high price of its first-generation product, Ceredase, had been the collection and processing of thousands of placentas to produce the drug. It was a time-consuming, complex, and expensive process. A higher price could be justified.

But many expected Cerezyme, made through a recombinant process, would be less expensive than its predecessor. Termeer concluded, however, that a lower price could not be justified given the expense of the new Allston facility, rising quality standards, and the superior safety profile of the new product.

"So Lisa calls me on the day they get Cerezyme's approval," recalled Abbey Meyers, "and she says she wants me to be the first to know about the approval. This was 1994. 'I want you to know that Henri told me to call you to tell you we're not gonna charge more for Cerezyme.'"

On receiving Termeer's dispatch, Meyers erupted, "What the hell are you talking about? You can make it in a big batch, you can treat thousands of patients now. It's not costing you for every placenta that you collect from a hospital. Why can't you send a signal to the community that you're going to lower the price?"

Raines would have none of it. Meyers recalled the conversation, "Raines replied 'Oh no. We can't do that because it costs so much for all the research to make this biotechnology product.'

"I was so angry. I still think about it. I think about how angry I was then. How could they have done that?"

Meyers was not alone.

Termeer later summarized the strategic rationale for Genzyme's pricing in an all-hands 1995 memo to the company's organization, "Critically, the revenue from mature products like Cerezyme allows us to continually innovate on behalf of other patients who need it. Our ability to develop breakthrough treatments for Fabry disease, MPS-1, and Pompe disease is directly tied to our success with Cerezyme. Our investment in Myozyme alone has already topped $500 million, and we expect it to grow by several hundred million in the coming years."

In the biotech world, a blockbuster was gold. Every company needed its winners to offset its losers. Cerezyme would be Genzyme's biggest winner, enabling the company to pursue other new therapies.

So as Clinton's healthcare reform package dissolved into the annals of U.S. legislative history, the biotech industry may have assumed it would have a respite of quietude before the next wave of legislative reform washed ashore. Nothing could have been further from the truth. From 1995 through the end of the next decade, healthcare and biotechnology reform remained front and center.

In 1996, Senator Ted Kennedy would rear his head and assume a more prominent role as one of the nation's top healthcare legislators. Kennedy was by no means a newcomer to healthcare. He had introduced his first national healthcare legislation in 1969, and he had taken an interest in the field ever since. As the "Lion of the Senate," however, he took charge of the national healthcare agenda as other legislators, notably Newt Gingrich, faltered.

For Termeer, this development was nothing short of providential. Senator Kennedy and he had gotten to know each other over the years. Although Termeer maintained good relations with leaders from both political parties, Kennedy was special. Both were a part of Boston's powerful elite, and both were outgoing men who enjoyed a good laugh. Furthermore, they were sailors, and Kennedy would occasionally invite Termeer for a cruise off Hyannis Port on his 50-foot wooden Concordia schooner, *Mya*. It was their ultimate release, although Kennedy never shut off the "Senator switch," using their time together to talk about policy ideas.

Kennedy knew how important Termeer had become as a voice for the global biotechnology industry. And conversely, Termeer recognized Kennedy for his leadership role in setting the U.S. political agenda, especially when it came to healthcare.

Michael Myers, who served for 23 years as a senior member of Senator Kennedy's staff, took over in 1997 as staff director of the Senate's powerful Committee on Health, Education, Labor, and Pensions. Chaired by Senator Kennedy, it was the committee through which all important legislation that pertained to the FDA and health regulatory matters flowed.

Myers remembered Termeer and Kennedy and the respect they held for one another, "When these two men got together, there was always electricity, a spark, in the room. They would usually meet in the Senator's Russell Building office, or sometimes up in Boston when the Senator was visiting town. Henri was an exciting emissary for the industry. Anytime there was an FDA issue that came up, Henri was one of the first people that Senator Kennedy would have me call, just to get his take on it.

"We did some pretty comprehensive stuff. Probably the most memorable thing we did was the biggest overhaul of the FDA statute in history. You can imagine we had a number of sticking points, but Senator Kennedy was always interested in finding a sweet spot in there somewhere.

"The length of product exclusivity was usually one that required a lot of discussion. The Senator wanted to make sure that we provided enough room so the industry could really go for the fences in developing blockbuster new drugs, but at the same time to make it short enough so that there was pressure on keeping prices down.

"Another thing we did was work on the creation of a regulatory pathway for the approval of biological generics. Henri was a critical player on this issue. Senator Kennedy was working with Orrin Hatch and we were trying to figure out, with this FDA bill, how to handle generics.

"Henri was able to convince us that there really is no such thing as a pure generic biologic. Henri even changed our vocabulary, and we came to call the generic versions of biologics, 'biosimilars.'"

Termeer's approach in working with the Senator was clearly one that resonated. Myers continued, "Henri's arguments were always forceful, but they were never strident or doctrinaire. He had this talent for really giving insights. He would do it with a twinkle in his eye. He would do it as a thoroughly knowledgeable person. He was very convincing.

"Senator Kennedy had enormous respect and affection for Henri. You could tell they liked being with each other. I think Henri saw the Senator as someone who not only was interesting as a person, but also someone who kind of got the vision and the potential of the whole bio industry.

"They were quite a pair."

One of the final tributes accorded to Termeer's courageous advocacy of his industry's policies came five months after his death. *The New England Journal of Medicine* would publish a paper that Termeer and Michael Rosenblatt, MD, an esteemed academic medical figure and former member of Merck & Co's executive committee, had coauthored. It was entitled, "Reframing the Conversation on Drug Pricing."

The authors expressed their minority view, counterposing the majority opinion offered by the other committee members with whom they had served. The National Academy of Sciences, Engineering, and Medicine committee had been convened earlier in the year to assess *Making*

Medicines Affordable; A National Imperative. The 235-page promulgation would be offered "In memory of Henri Termeer, 1946–2017."

Dating from his installation in 2005 as BIO's President & CEO, Jim Greenwood had known and worked with Henri Termeer for 12 years. This highly respected leader of the biotech industry's most powerful, international trade organization had also, prior to his appointment at BIO, served as a social worker, state legislator, and six-term member of the U.S. House of Representatives (PA-8).

Greenwood would remember Termeer in this way, "Henri was the personification of an innovator of a high-priced drug. Because of the pricing of some drugs, a lot of people hate our industry. It's a great irony because we save lives and we improve lives, and everyone in the industry is really dedicated to that. But then you have these high prices and this perception that this industry is basically populated by greedy people who take whatever NIH does, tweak it a little bit, and then start gouging patients.

"Yet Henri was so patient-focused and had such a dear heart ... just a loving, good man ... nobody who could listen to Henri speak or speak with him, whether it was a member of Congress, or a reporter, or anyone else, could come away thinking that he was doing anything except the Lord's work."

CHAPTER EIGHT

From Brazil to China

Uzma Shah, MD, stood quietly in a corridor on the fifth floor of a major regional hospital in Karachi, Pakistan. It was 1995, at the height of Operation Enduring Freedom, the U.S. and allied invasion of the Taliban in Afghanistan. Thousands of refugees had streamed across the border, seeking shelter. The ward was filled with patients in distress—lost limbs, blindness, burned bodies, infections, and other trauma. Amid the heat and chaos, she pondered the plight of one particularly critically ill five-year-old boy.

He was in pain with difficulty breathing, near cardiac failure. The diminutive Dr. Shah, 33, was visiting her native Pakistan for a week from Boston where she had recently completed her medical training at Massachusetts General Hospital (MGH). Shah's parents had moved from Pakistan to Surrey, England when she was two, but they had always wanted her to return to Pakistan to do good, or as she would later describe it, "to do my job for humanity."

Her Division Chief at MGH had suggested she visit Karachi; the hospital was looking to add a pediatric gastroenterologist to its staff. Now, she wandered the halls of the wards, taking in the new environment, opening her mind and her heart to the possibilities before her. In one of the wards, she noticed a child who was in obvious distress. She found the residents and the medical team charged with caring for the young patient, and they told her he had Gaucher disease and was in cardiac failure, renal failure, respiratory failure—his little body was shutting down, organ by organ. His diagnosis was so dire they had put a Do Not Resuscitate order on his chart. There was nothing they could do to save his life.

"It was very hard for me to just stand by," Dr. Shah recalls. "I spoke to the team and I spoke with the parents. I asked them to allow us to revisit his situation and see if we could do something else.

"I don't know how we did it, but somehow we managed to bring him out of cardiac failure," Shah remembers. "We resuscitated him and he survived.

"It so happened that a week later, I was in Boston having dinner at a friend's house. I mentioned that a Pakistani hospital was allowing me to see things I had never seen before and situations that tested not only my abilities as a physician, but more importantly tested my sensibilities as a human being.

"I shared that there was this five-year-old who was really worrying me, and my friend, who was working at Genzyme said 'wait a minute, I might have somebody who can help. Her name is Tomye Tierney.'"

Dr. Shah was skeptical. There was silence at the dinner table and she looked at her friend and asked what would be the point? The child in Pakistan was dying. There was no hope—or was there?

Tomye Tierney was Genzyme's liaison to the developing world. She was tasked, among other things, with serving patients in lesser developed countries. It was a role that emphasized Genzyme's commitment to treat those in poor countries that were unable to cover the costs of expensive drug therapies.

Cerezyme had only recently just been approved, and while Shah was aware of its efficacy in treating Gaucher patients, she knew only too well that it was beyond the means of Pakistan to provide it to those in need. But Genzyme had a whole division devoted to humanitarian interventions. What other drug company did that? Maybe there was hope for this little boy in Pakistan after all.

A few weeks after Shah returned to Boston, she got an e-mail from Tomye Tierney that said Tierney was intrigued by this case and she wanted to help. Could she come with her and meet Henri Termeer?

The next day, she was escorted up to the 12th floor of the company's beautiful, glass headquarters.

"I met Tomye," Shah recalls, "and we hit it off immediately. She said, 'I'd like you to meet Henri. Henri wants to meet you.'" Getting off the elevator, she and Tomye walked down to the corner office where, as usual, the door was open. As they approached, there was Henri who hopped up to greet her and inquired almost reflexively, "Can you please tell me about this young boy?"

Uzma Shah answered, "Well, last time I saw him, his heart is ticking, his kidneys are functioning. He's enjoying his breakfast and his lunch and his dinner. His parents are in a refugee camp and I don't have the heart to see him just go. This is a beautiful boy."

When she was finished, Henri said, "I'm going to make you a promise. The promise is as long as I am alive, this child will get treatment free of charge. This is my promise to you."

Shah could not believe it. Her eyes welled with tears. Henri Termeer saw her reaction and moved closer to her. "Come and sit down with me because I think we're going to have to cry together."

Shah was overjoyed but skeptical. "In my heart I was thinking, 'Well, you might change your mind because this is very expensive ... this is Pakistan we're talking about. There's a war going on in Afghanistan. They live in a refugee camp. How on earth am I going to give him this injection once a month, get it through immigration and get it to this kid?'"

Termeer—whose watchword to his team was always "it's your responsibility" —said, "That's your job. I'm going to promise to get it to you."

Shah couldn't believe it. Leaving Henri's office, she turned to Tierney and said, "He doesn't mean that, does he?" Tierney said he did.

"He means it. He's already told me this. He's instructed that I better not drop the ball, and, Uzma, I'm not going to drop the ball." It was her responsibility too.

Several weeks later, Shah returned to Karachi. To her amazement, the medication had been delivered by Pakistan's Ministry of Health. It was an ordeal, but Henri and Tomye had kept their promise. Shah followed this child for years. He received his regular injections. He survived. Shah described him as "a beautiful young man."

The hospital kept its pledge as well, that as long as they received the drug, they would ensure that this child would get Cerezyme free of charge.

The years went by. Every few months, Tierney would write to ask how the young boy was doing and to check on his medicine deliveries. The family would take pictures and send notes of thanks for "basically saving his life." But they didn't really understand why this miracle was happening to them.

Shah remembers the boy's mother asking of Henri Termeer, "But why does he want to help us? Why does this gentleman want to give us this for free for the rest of his life when I know this is very expensive? Do they want something?"

Uzma Shah said, "No, they don't want anything. He wants nothing from you.

"In my books," says Dr. Shah, "there are not very many people like that in the world." Dressed in her white physician's coat, with her mane of curly, dark hair and deep brown, soulful eyes, Shah concluded, "That was Henri for me."

Uzma Shah felt that her experience with Genzyme was remarkable, unprecedented even, but that kind of humanitarian commitment was the foundation on which the corporation was built.

Stories of this sort were repeated around the globe time and again as Genzyme grew into a Fortune 500 company by 2009. The mission alignment became Genzyme's signature and created its own kinetic energy among its employees at all levels. It also served to brand the company with its patients, their families, physicians, and others concerned with the treatment of rare disease patients.

Genzyme reflected Henri Termeer's most fundamental, singular, and enduring qualities. Henri Termeer was an optimist. Henri Termeer was also a pragmatist. And everything Henri Termeer did was imbued with his unshakable spirit of hope.

For Genzyme, that meant pioneering and discovering a new therapy was not enough. Patients who needed it had to be able to access it.

Sandy Smith, former head of Genzyme International and a 15-year Termeer colleague, remembers those days, "In the earliest days, we really measured the business by how many patients we were treating. I was there from '96 till the Sanofi acquisition. The magic of the way we ran the company will maybe never be reproduced. The decisions that were made in the stairway at 500 Kendall, ... you'd run into somebody up and down the stairs or in the corridors and a decision would be taken and you'd move on. There was not the sort of organizational verticality to which I had become accustomed at my prior company, a Big Pharma. We managed the business by how many patients we were treating. Patients were at the heart of every decision."

In 1992, Ceredase reached Brazil, among the first developing nations to receive the new medicine. Rogerio Vivaldi, MD, an endocrinologist in Rio de Janeiro, was fresh out of medical school, establishing his practice. A professor asked him to see a 14-year-old boy who had been diagnosed with Gaucher disease by Robert Desnick, MD, PhD, a pioneering geneticist at Mt. Sinai Medical Center in New York.

In New York, the boy had been on a drug protocol that required administration of Ceredase through an intravenous drip every two weeks. That in itself sparked Vivaldi's interest. "I definitely want to see a patient that I will have to see every two weeks, that is all good."

The young boy, Alberto Levy, loved soccer but was susceptible to injury. He was going to the hospital a lot, bleeding often with an oversized liver and enlarged spleen.

As the treatment regimen began, Vivaldi established a routine informed by his own experience in intensive care. He was the one finding the boy's vein and setting up the IV.

"Some people would describe that as the moment when I saved his life," Dr. Vivaldi would recall, "but I describe it in a different way. It was a moment that two lives were saved. Not only his life, but my life as well. Because I saw something in medicine that we are always looking for. We are looking for a miracle. We are looking for something that is so effective that it blows us away. And that is what I saw."

The therapy allowed the young boy's genetic machinery to restore balance. "He grew almost 51 centimeters. He became taller than his cousins. He went back to playing soccer."

At one of their first dinners together, Henri Termeer asked how many Gaucher patients there were in Brazil. At least 200, Vivaldi replied. But as word of the treatment spread, and new patients were identified, the total number rose to more than 1,300, one of the largest concentrations of Gaucher patients in the world. Within two years, Rogerio Vivaldi had diagnosed more Gaucher patients than almost anyone in medicine.

But that was not enough. Brazil's Gaucher patients needed more than Genzyme's therapy and leadership. The government of Brazil would have to play a role too. The country's national healthcare infrastructure needed to be adjusted to diagnose and treat Gaucher patients. Vivaldi knew that a payment scheme needed to be put in place.

"I wrote a letter to the Ministry of Health in Brazil, as a physician, describing the disease, describing the patients, describing what I was seeing with Alberto, and the Minister of Health answered me back, saying, 'I like what you wrote. I'm reimbursing this drug.'"

Vivaldi called Genzyme and told Termeer he had the reimbursement. That was great news, but Henri, as always, wanted more. Now that he had seen what Rogerio Vivaldi was capable of, he wanted to bring him onto the team. Tomye Tierney was tasked with recruiting him, but Vivaldi did not want a job where his compensation was based on sales or how many patients he treated. "If you want to hire me, you hire me and I have a salary," Vivaldi told Henri Termeer. "That's it. I'm not going to be compensated proportionate to anything at all."

Henri's response? "Okay, join us. Create a company in Brazil."

It was blue-sky thinking at its finest—Henri Termeer style.

"They had no idea what we could do," Vivaldi says. "I probably had no idea, but I wanted to do a different pharma company. I was very critical of the sales reps visiting me in the office. I wanted to recruit the best people to do the best patient-centric thing, and that's what we did.

"We started a company with zero revenue, and in less than five or six years, we were selling more than $100 million. I think I got the largest contract ever signed by Genzyme by a single country."

Both sides came out as winners on the deal. The contract with the Ministry of Health was the largest contract in the history of that department.

"We had three different political parties, multiple Ministers of Health, different presidents, and we never allowed the contract to lapse," Vivaldi says with pride. The project in Brazil became the prototype on which Genzyme's other emerging market programs would be based.

Vivaldi's creative and passionate leadership led to his promotion as Genzyme's first head of Latin America where he oversaw implementation of the company's "One Price or Free" pricing strategy in the region.

This novel approach to pricing prescription, high-priced medicines held up for a period of time, but eventually it gave way to a more flexible approach as Sandy Smith described, "We did not have uniform prices around the world as much as we like to believe we did. It never really happened.

"Genzyme had an enviable compassionate use program. Then there were countries where we would negotiate. This is something that is still not understood today. We'd say 'Look, you have so many patients that we've been able to diagnose through your medical system. How about if we treat all the patients?' No two countries were alike. The other factor was that there were so few countries where you had more than 200 patients."

Under Vivaldi's leadership, Genzyme entered 16 Latin American countries, and in most, only two or three patients needed to be identified to establish the footprint on which infrastructure, training, and patient identification and recruitment could be built. If local governments were unable to pay for the drug, Genzyme provided it for free.

Rogerio Vivaldi understood the math this way, "At $300,000, if you give a 10% or 20% discount it doesn't change a thing. It doesn't make the drug affordable. The only thing that will make a difference is to give the drug for free.

"Are these patients at risk, medically speaking? If they are, the company has a social responsibility because the companies elect to play in the rare disease sector, with the understanding that not everybody can pay."

Vivaldi recalls taking Henri Termeer to a hospital in Rio de Janeiro where they visited with some of the youngest Gaucher patients, including a one-year old girl.

"Henri had tears in his eyes," Vivaldi remembers. "He was so happy that he said, 'Rogerio, to be treating this one-year-old girl... . This girl will never feel that she's a patient. She's simply too young to recognize the disease, and we are changing her life from what it was like before. That's the gold. That's the gold.'"

Over time, as Cerezyme's efficacy became clear, governments realized that it would become their responsibility to reimburse for the lifesaving therapy. As Termeer explained, "Our goal must be to create a situation in which the country itself will eventually take responsibility for the treatment. That's where we need to get to."

The Genzyme success story in Brazil and, eventually, throughout Latin America, combined with Termeer's compassion in addressing the Gaucher community's needs, led to the establishment of Genzyme's Gaucher Global Initiative in 1998.

The GGI, as the initiative came to be known, quickly linked itself in 1999 with Project HOPE (Health Opportunities for People Everywhere), a global humanitarian provider of health solutions for patients across five continents.

Together Genzyme and Project HOPE established an innovative partnership to provide enzyme replacement therapy for Gaucher disease patients. In its first five years, more than 200 patients in 17 countries received Cerezyme through the GGI/Project HOPE alliance.

Termeer often summarized Genzyme's involvement in such a venture as the company's "responsibility." It was Genzyme's "obligation" to treat these patients, whether they could pay for it or not.

The initiative between Project HOPE and Genzyme was vintage Termeer. It was global. It was visionary. It was pioneering. It was intended to serve those who had no hope, no access to therapy, and little or no financial resources. It was another illustration of the high moral standard that Genzyme would seek to set in making sure that those in need of treatment would receive it.

John Howe, MD, who led Project HOPE as CEO from 2001 to 2015, described Termeer's partnership in the venture with a phrase he often heard Termeer use.

"When I would visit with him he talked about the importance of, in his words, the portfolio approach. We talked about it in the Yellow House and it wasn't about his portfolio, in the traditional sense, of here's the capital to be allocated, and then you put the pie together and the investment occurs.

"Mainly, this was about a portfolio approach to his time. Later, it would be about his time with his partners, his time at Massachusetts General Hospital, his time at MIT, his time on boards in biomed, but here he had a portfolio approach about how to allocate time and resources for treating patients.

"He asked, 'how do you put the puzzle pieces together to execute a moral responsibility? How do you put the puzzle pieces together to get the medicine to those in need?'

"Genzyme would provide the medicine and Project HOPE would then deliver the medicines, create an in-country network, train physicians, attract patients, build capacity, and develop the health infrastructure.

"The picture I'm painting is that it wasn't a one-off. His hope and aspirations were that patients in need would get the medicine but we'd do it by not just showing up and leaving it at the doctor's office, because the physicians in Egypt and China had no experience in administering the drug. Our responsibility was also to train the doctors and very importantly to develop the health infrastructure."

Pramod Mistry, MD, a Professor at the Yale School of Medicine who was a leader of the Global Gaucher initiative, described Egypt as a spectacular story.

"You can just imagine Cairo. It's a very chaotic environment. When we started, nobody knew anything about Gaucher disease. And then in the early days, Rich Moscicki (Genzyme's first Chief Medical Officer), Tomye Tierney, and I would visit, hold clinics with interested physicians and see patients, examine them, and discuss therapy options. At the time, Egypt had no healthcare system for the treatment of rare genetic disorders.

"We became a platform to mentor a generation of Egyptian physicians across the whole of Egypt who eventually assumed responsibility for the country's Gaucher patients. They are now experts in Gaucher disease and they possess a level of sophistication that rivals that found in the most advanced Western medical communities. We were urged to treat this group in the same manner as if we were treating patients in Manhattan."

John Howe credits the wife of Egyptian President Hosni Mubarak for supporting Project HOPE in a difficult environment. "Mrs. Mubarak was a great champion. When there was upset in Egypt, and things were unsettled periodically, the ship would come in with the medicines and they would put a hold on delivery of goods, but she was always helpful in finding a way to get the ship with the Cerezyme in so the patients would have it."

In 2015, signaling the importance and success of the GGI, the Egyptian government assumed financial responsibility for funding Gaucher patient therapy and for funding the nation's Gaucher patient centers.

This humanitarian program was replicated in China, India, and various African countries. It was a bold initiative. Many obstacles were overcome, ushering in the next era of rare disease treatment for all patients worldwide regardless of their means. As Termeer often said, "Our goal is to serve the seven billion people of the world, not just the one billion fortunate to live in developed countries."

In China, Termeer installed another promising young leader, James Xue, as general manager of Genzyme's China business. Xue, a U.S. citizen born in China, had first met Henri Termeer as a business development intern at Genzyme from Henri's alma mater, the Darden School of Business. The two just clicked.

Termeer told him, "James, you have a unique competitive advantage. You speak the language, you understand the culture, and China must be one of the most important markets for Genzyme."

Termeer harkened back beyond Darden, to Baxter. "I think one of the best ways you can develop is to do something very similar to what I did at Baxter," he told the young man. "I spoke German, I understood Europe. I was able to really build my career foundation by being in Europe."

But while being groomed, before he made one of his first Genzyme trips to China in 2003, Xue remembers a pivotal meeting in Henri's office.

"A day before I left, he invited me to his office," Xue recalls. "He tried to convince me that we had exceeded the number of patients who could be treated free of charge. He said that's not sustainable. He said, 'China's responsibilities are to take care of their own patients. We as an industry, as a company, we can only do so much for so long, right?'"

But James Xue didn't reflexively answer, "Right."

"Without too much thinking I said, 'well, it's going to be very hard to make reimbursement happen in the Chinese cultural context ... to really allow the society to pay so much for so few patients."

Termeer's reaction shocked James Xue. "That could have been the termination of my career," he thought at the time. "His face turned red and he gave me one of the most memorable lessons in my adult life. He said, 'James, if you do not believe in this, you should not go to China. You have to be a believer. Any society, any market, has to take care of its patients, the people who are not in the position to protect themselves. That's part of what we do, and you have to be a believer before you become a successful practitioner.'"

Years later, Henri would apologize—more than once—for his harsh tone during that meeting. "'James, I was too tough on you,'" Xue recalls him saying. "But I said 'Well, Henri, that was really one of the best, most crucial times you have given to me.'

"And of course, he later saw how I became transformed into not only a believer, but a preacher, a mentor. I almost use the same line today for the executives that I have mentored at my new company, CANbridge."

Support for all these investments in developing countries came in part from the payoff from years in bringing large wealthy countries into the field.

Japan, a market that Genzyme entered in 1987 through the sale of diagnostics products, was a slow adopter of Ceredase. Despite a large population of Gaucher patients, its Ministry of Health and Welfare (MHW) would not permit Genzyme to introduce the new drug for many years after its approval in the United States in 1991.

MHW claimed that it was too risky, relying on human tissue (placentas) as a source of drug product that might be HIV contaminated. Termeer was aware that there were at least 36 children in Japan who were afflicted with Gaucher disease and needed treatment immediately. He and Genzyme's country manager went to the then–U.S. Ambassador to Japan,

Walter Mondale, to discuss ways to provide access to the drug for these sick kids.

Their strategy was bold. Termeer suggested, "Let's move the children to Hawaii and treat them there," and Mondale was supportive. Once the Japanese government officials were made aware of this potentially face-losing stratagem, they reversed course and moved quickly to approve the product.

Once again, Henri Termeer was demonstrating Genzyme's "prime directive" of responsibility to patients. This responsibility was, according to Henri Termeer, "... the backbone of everything we do. If you interviewed the 11,000 people employed at Genzyme, that's what gets repeated, because that is our purpose. It is bigger than any one of us individually. It is the current that drives us and it's stronger than stock prices, or this, that, and the other."

The European region, Genzyme's largest international market, posed a challenge of a different, but related sort—the intransigence of certain government bureaucracies to enable the introduction of an expensive new drug therapy. In the United Kingdom, NICE (the National Institute for Clinical Excellence), the body that rations the nation's drug and healthcare spending, declared that it was too expensive to reimburse and that the government could not afford it. Whereas their position was not the only challenging one in Europe, it was among the least accommodating. The Genzyme team would work for years with patients and physicians to overcome it.

The Nordic region provided, however, an illustration of markets in which Genzyme faced less severe challenges. Dick Meijer, who joined in 1991 as Genzyme's second employee in Europe, had the assignment to create patient access in six northern European countries.

Sweden was among his first priorities. Working in a remote area near the Arctic Circle, Meijer met a clinical expert, Dr. Anders Erikson, who was the leading Gaucher physician in the Norrbotnian region.

One of Erikson's patients, a young boy, was wheelchair bound and the first patient in the region, among the 25 who had been diagnosed with Gaucher disease, to be treated with Ceredase. The authorities granted approval for reimbursement instantly. Although they challenged, on the basis of price, the conversion of patients from Ceredase to Cerezyme, they continued to reimburse the annual treatment costs for both drugs.

Meijer recalled the young Swede and the overall receptivity of Gaucher therapy in Norrbotten, "The little boy with a big belly started taking

Ceredase. His recovery was remarkable. He was not splenectomized and in five years he was playing ice hockey. All the patients got treated and the government reimbursed them. This movement occurred in other Nordic countries as well."

Sweden's experience was similar to certain other large, developed, wealthy nations. The formula that Termeer relied upon in building the company's international markets was proven to be straightforward, if not obvious or simple to implement.

Jim Geraghty, who led Genzyme Europe starting in 1998, saw success there traceable to the roots of its pioneers, "The first people to introduce Ceredase in Europe, like Jan van Heek, Carlo Incerti, and François Cornu, had all worked at Baxter, as had I. The fact that we knew Henri's values, and shared the same management philosophy, created a trust that allowed the European team to operate very entrepreneurially."

He would hire many physicians to build these markets. Most of these leaders would be nationals. Patients would lie at the center of every single thing that was done to build the franchise. And acceptance of individual responsibility was at the core of all decision-making.

Termeer cared deeply and personally for his country manager/leaders. He believed in seeing them frequently. He and his colleague, Sandy Smith, would periodically fly a chartered Gulfstream VII around the world out of Hanscom Air Force Base and visit ten to fifteen countries in a week.

Annually, he would also invite the entire group of international managers to his home in Marblehead for a festive dinner. Part of the ritual for these Genzyme gatherings was the salutary speech that each country leader was expected to make to the assembled group, reinforcing the cultural vibe pulsating through the room. New hires were inculcated, often peppered, with some good-natured ribbing, as they would be asked to stand on their chairs and deliver remarks on their decision to join Genzyme and the challenges they faced. These extraordinary, celebratory events were essential to building the global connectivity and Genzyme culture.

But in the end, it all came back to doing the right thing for patients, individual accountability, and personal responsibility. These three pillars were the basis of Termeer's leadership credo and his contract with the entire organization.

CHAPTER NINE

HAT Tips

Henri A. Termeer loathed strategic planning. He had been asked many times to conform to the consultants' models of formalized mission statement debating, long-term strategizing, and five-year outlooks. Each time, he bristled. What Henri wanted was "strategic thinking."

"There was a joke about Henri that he didn't like the word 'planning,'" remembers Greg Phelps, one of Termeer's closest friends and a long-time colleague at both Baxter and Genzyme. "He didn't want a strategic plan, because he was afraid that a plan would encumber people's range of thinking."

Henri would often refer to the way long-term strategic planning had been done at Baxter as an example to be avoided at all costs. What a waste it was to essentially remove top executives from the field of play for a month, just to create reams of paper for three- to five-year plans that would be put in binders, placed on a shelf, and forgotten about. It was a mechanical process that prevented what Henri valued most—strategic thinking, flexibility, and agility in fast-evolving markets.

"We were forbidden from even talking about long-term strategic plans," Sandy Smith recounted. "One guy from one of the big financial houses tried to do it," Smith laughed, "and he was attacked. He had to leave."

Henri Termeer believed in being opportunistic, and strategic plans got in the way of that. The more formalized a plan you had, the more it interfered with communication. You wanted to be nimble and opportunistic, he would say, not bureaucratic.

Henri Termeer's definition of opportunism began by defining the mission. "He was your business partner," Greg Phelps says. "Whatever your role in the company was, if you knew what the mission was, and you were a sharp person, and stayed close to Henri so he knew what you were doing, he was behind it."

Although Henri relied on some scheduled meetings, he kept much of his schedule open for casual drop-bys and meetings on the fly. Structure was the enemy of opportunism. If you were taking your cues from a rigid notion of structure, you were going to spend your days living in that box.

There were some structures, however, that Henri could not discard, nor did he want to. What Henri did embrace was a tightly focused process. The weekly review meetings—attended by the roughly 30 persons who had Profit & Loss responsibility—took place every Monday morning at 7:00 a.m. in Henri's large conference room on the 11th floor, a flight of back stairs below his office. As his team assembled for these sessions, if someone new to the group tried to sit in the chair at the left end of the horseshoe table, they'd be warned off—that was Henri's chair.

Jan van Heek, a Dutchman who served at Genzyme in senior international leadership roles from 1992 through the company's sale in 2011, remembers how he used to feel on Sunday nights, "I can tell you, Sunday nights were sometimes tough nights because Monday morning you had to stand up and talk about issues, talk about numbers. When things weren't going well, it wasn't fun standing up."

In fact, many spent Sunday nights calling Termeer to give him a heads-up of bad news they would be presenting the next morning. Henri did not like surprises, and they knew how important it was that he not be blindsided in a group setting.

"Okay, where are your numbers?" he'd ask each member of his team. "What did you expect? What happened? Are you changing your forecasts?"

Given the focus on public company investors, Sandy Smith remembers, "Patients were at the center of everything we did. But we were very, very conscious of EPS (earnings per share)."

Termeer would traverse the globe in these sessions, peppering his managers with questions about how Genzyme was doing in international markets. "What happened here?" he would ask. "Are you working on a solution?"

No one had a printed agenda, but everyone always knew how the meetings would transpire. For the next hour or hour and a half, each unit head would give an update and go over the numbers. And when it came to the numbers, no one ran a tighter ship. Termeer had a knack for finding the one number on the slide that did not add up. Someone would be presenting a slide when all of a sudden Henri's green laser pointer would zero in on a line item.

Everyone knew about Henri's green laser. "It became legendary," says Alison Lawton, a 20-year executive at Genzyme, having joined the company in 1991.

"Most laser pointers are red, but Henri, the eternal optimist, loved the fact that his was green."

Sometimes, though, the last thing you would want to see was Henri's green laser, symbol of optimism or not.

Alison Lawton recalls, "If you were presenting slides, one after another, packed full of numbers, within seconds Henri, with his green laser pointer, would find the one number you didn't want him to find and he'd ask you a question and he would see it just like that."

And he had an exceptional memory. He astounded Genzyme's leaders with his ability to remember the most obscure, minute detail from previous presentations made months earlier. You could call it photographic recall. You had to be prepared, *really prepared*, for these meetings.

But the meetings were a positive force within Genzyme, providing the team an up-close, unvarnished view of what everyone was doing or had to do. After all, it was their responsibility.

No one felt that Termeer was trying to trip up his managers. Alicia Secor, one of those who attended regularly and was one of Termeer's mentees, said, "He demonstrated that he was willing to take the time to show that he was paying attention and that he cared. He would also point out things that we might not think would be obvious to him. So in that sense it was almost like a gotcha moment. But what he was really saying is that I ought to be focusing on this or that and you would need to be prepared to answer to him... . It was a tricky balance between focusing on the numbers, the performance, and making people feel motivated to continue to do what they did. He would push people hard, but always in a fair, constructive way. He could really challenge you but you could feel good afterwards, and that was one of his special skills."

Termeer was a man driven more by instinct, comfortable with ambiguity, and governed by a sense of decency and civility. But he was also, in some ways, a contradiction. Outwardly he could be and often was a serious, ambitious, competitive businessman. But he was also a flexible, deeply compassionate, fun-loving, and curious man who enjoyed people, generously shared credit for success, and eschewed structure or rigidity. At his core, his moral values and ethical principles were nonnegotiable.

There were times he could be stubborn. And he clearly liked to be in charge. But he was just as comfortable providing others with direction, holding them accountable, and stepping back, out of their way, giving them ownership of the matter at hand.

"Henri was very happy to have different ideas and different perspectives," Alison Lawton says. "He would go around the table and ask every individual, 'What do you think? What do you think? What do you think?' He wanted to force everybody to verbalize what they thought so he could understand their perspectives. But he could also react. If somebody challenged him inappropriately he would stand up to that and he would not appreciate it."

Termeer was an optimistic man. He was often seen smiling. Many remember him just this way. Yes, he could be tough, very tough. And in his earlier years at Genzyme, his temper was on a rare occasion revealed. But as he aged and gathered stature, power, and accolades, he was increasingly comfortable in his own skin. He was a sensitive man in control of himself—humble, human, and grounded.

Intentionally locating his office by the entrance to the employee cafeteria on Genzyme Center's 12th floor, he was visible and highly accessible to all of Genzyme. It influenced how he was perceived by others, lived his life, led the organization, and approached decisions. He was a larger than life figure who brought his considerable talents, charisma, and intellect to everything he did.

To Joan Wood, who led organizational development at Genzyme for many years, Henri brought the same informal approach to developing people as to strategy. He never liked org charts or wanted formal mentoring programs. He believed in the power of the human connection to develop a team of inspired leaders. Wood recalls, "Henri had a personal force of energy that made it impossible for even the shyest among us not to connect."

Perhaps his closest relationship at Genzyme was with long-time confidante and Genzyme executive, Peter Wirth. "Henri's defining characteristic was his natural, instinctive ability to empathize and genuinely engage with people," says Wirth. "Whether he was talking to a patient about their desperate need for access to an effective drug to treat an illness that affected only one person in 40,000 or sitting down in the Genzyme cafeteria with a group of unsuspecting young MBAs for an impromptu review of their latest corporate development project ... he was unselfconsciously but intensely present... . You had the feeling at that moment in time that you were the most important person in Henri's life, and that feeling somehow left you with a sense of comfort, confidence, and renewed optimism."

At his core, Henri was a wise enabler. He selected enormously talented people, aligned them to the mission of the enterprise, coached and mentored them, and watched them grow, deliver, and self-actualize. He was not one to hover. Rather, he would err, if anything, on giving others too much room. The burden was transferred to his team. But the ultimate responsibility, at least at Genzyme, rested with him. The crisis at Allston Landing was a clear illustration (see Chapter 10). The buck would stop at his desk. And as it did, he would accept full responsibility for the plant's viral contamination and the fallout that ensued.

When asked if Henri was a genius, Victor Dzau, MD, long-time Genzyme Board Director and President of the National Academy of Medicine, described him this way, "Einstein's a genius. Henri was a leader in the best sense of the word because he understood people who worked for him and people who actually supported him, including those who bought his products. He was a leader because he was bold and willing to take risk. He was a leader because he was decisive and made things happen. He transformed an industry and did it in such a way that it was almost natural."

Or, as Jack Heffernan puts it, "Henri was a father to all of us as we were his children. We were family. You'd get admonished and scolded and then you'd get hugged. He encouraged you. You talk about ownership. Henri would say, 'You own this. It's your job, not mine. You need to take care of it' ... But he'd never let them fail. Henri always had a net there... . He would take the Genzyme culture net and spread it like a Greek fisherman... . You had to be secure that people weren't trying to undermine you. That goes back to loyalty and trust. Optimism is a beautiful thing, especially when it benefits the patient. There was constant inspiration. In the end, it was our challenge and our goal to save the precious lives of patients."

You had to be a strong executive to take on the boss, but those who knew their stuff usually, in the end, benefited. They would earn Henri's trust and respect. Biochemical geneticist Ed Kaye, MD, was one such leader. Ed had joined Genzyme in 2001 from the prestigious Children's Hospital of Philadelphia. He was widely recognized as an international key opinion leader, or KOL as they are known, in the field of lysosomal storage disorders.

Dr. Kaye described a meeting of about 25 executives early in his nine and a half years at Genzyme.

"One of the things I found interesting was that people were so loyal and respectful of Henri. Sometimes they had a hard time disagreeing with him. The first time that we had a disagreement, it was on the design of a very large clinical study I proposed for the use of Myozyme in adults with Pompe disease. Henri just thought it was a dumb idea. 'Just do it in kids, show it works, and that's it.'"

But Kaye felt that Termeer's view was not taking proper account of a major detail. About 80% of the patients afflicted with Pompe disease were adults. Kaye would not back down. He said, "Henri, you do know it's customary to actually study the drug in most of the people who you are going to treat?"

Termeer could be stubborn, but what Kaye later realized is that he was attending his first "HAT summit," the playful shorthand his team would use in naming these colloquia for their leader. He explained, "HAT summits were basically all the senior execs kind of discussing a problem, and Henri would be there. And I remember trying to have to defend why I wanted to do this very large, expensive, Phase III study in Pompe disease in adults. It was like a Sam Peckinpah movie where everyone gets shot in three-quarter time. And I'm up there trying to defend it. Henri was mad, and he was mad at me.

"But when somebody goes after me, and I think I'm right, I mean, you'd have to take a bazooka to get me to back down. It's just my nature. And so I'm sitting there arguing with Henri. I think we were kind of shouting. Nobody was speaking, and *everybody* had agreed with me before. But I was the spokesperson for the group."

After a heated exchange, Henri turned to another meeting attendee to gather another perspective on the Pompe disease patient community. She said Kaye was right. The meeting had climaxed. A half hour had passed. Termeer relented, "All right, all right, I'll let you do it. You're excused now."

After these arguments, Termeer was known to follow up, make amends, and clear the air. A few hours later, after the meeting had concluded and the heat dissipated, he would appear at Kaye's office door, offering a reconciliatory olive branch.

"Henri, am I fired?" Kaye would sheepishly ask in welcoming him. Termeer reflexively responded, "Of course you're not fired. I need to understand why you're so stubborn about this." It was a classic Termeer moment. Kaye would later laugh, "Can you imagine? Why *I* was so stubborn?"

Kaye went on, "So I explained it to him. He allowed that intellectual honesty to occur."

Henri's Dozen

Although Henri never was one to formulate a formal set of rules, he embodied and lived by a set of principles that those who worked with him would recognize as core to his leadership style. From many conversations, a dozen central themes emerged as those for which he would be remembered most frequently.

1. *Patients Always Come First.* Henri Termeer was "patient-centric" before it was cool or the phrase had even been coined. If it was the right thing for the patient, Termeer encouraged his colleagues to challenge the system, "We work in this space by permission of the regulators, patients, and physicians, and we do it with humility."

2. *Exalt Innovation, Especially If It Saves Lives.* Henri Termeer's devotion and love for innovating, beating the odds, and achieving the unachievable were core to his appeal and success. The development of Ceredase, a program on which Termeer bet the company, will be remembered forever as one of the gutsiest decisions ever made in biotech's storied history. His scientific advisory board recommended he abort it. He had said to them, "You may be right but we are going to pursue it anyway." The consequences of this bold decision reverberate to this day.

3. *Be Accountable; Operate with Integrity; Take Failure Personally.* Termeer was devastated by the 2009–2010 breakdown of Genzyme's flagship production site and the supply outages suffered by his beloved patients. He felt personally responsible. "It's your responsibility," he

often told his colleagues and mentees. "This sense of responsibility is the backbone of everything we do at Genzyme. This is our purpose. It is bigger than any one of us individually. It is the current that drives us."

4. *Take Risks with People; Challenge Them; Let Them Run; but Do Not Let Them Fail.* Termeer was a brilliant human capitalist. He was a master mentor who knew how to attract and get the most out of his enormous base of talent. He supported his people, nurtured them, taught them, genuinely cared about them, and built a culture that embraced them, rooted in Genzyme's humanitarian mission.

5. *Strategic Thinking, Not Strategic Planning.* Termeer was famous for his abhorrence of strategic planning. His riposte was "strategic thinking," an approach he had learned at Baxter. He hesitated in accepting detailed forecasts beyond 18–24 months. He was highly opportunistic, and he believed that an overreliance on long-term planning could be paralyzing. Although his mother had made him give up his chessboard, he was still the consummate chess player who constantly looked at the board, evaluating a dynamic, ever-changing set of options. He was usually four moves ahead of everybody else in the room.

6. *Be Mindful of Your Time.* Henri liked to tell the story of how his father had given him his first watch on the occasion of his 12th birthday. Bo Piela, Genzyme's communications director, recalled Termeer's words, "'I remember sitting on my father's knee. I put the watch on. My father pointed to the second hand. 'Do you see that?' he asked. Every second that passes never comes back.'" The speed of time created a sense of urgency within Termeer's soul. It constantly fired his passion for saving patients' lives. He collected clocks. They occupied shelves and shelves of space at his Genzyme office and, later, at the Yellow House. He was said to be a fast walker, a busy man with a pace, who always made time for people.

7. *Disdain Bureaucracy; Tolerate Ambiguity.* Disciplined but often freewheeling at the same time, Termeer avoided structure that would hinder crisp decision-making. At the height of Genzyme's growth, he had as many as 30 direct reports. He was always the hub, and the

organization its spokes. He never relied much on organization charts, and neither did those who surrounded him. If he knew he could find it, he thought nothing of dropping four layers deep, suddenly appearing in a junior colleague's doorway looking for answers.

8. *Celebrate Heroes, Especially Those Who Take Action to Improve Patients' Lives.* Termeer recognized the critical importance of commending those individuals and teams who achieved great things, especially those aligned with Genzyme's patient-driven mission. The annual Alpine Awards were established to anchor this cultural imperative.

9. *Do Not Give Up Your Firstborn.* Termeer was unlike most other biotech CEOs of his generation. They often partnered their first, most prized, and valuable R&D programs. It was their way to finance their business's growth. But Henri Termeer would not relinquish control over Genzyme's destiny, and so, among many alternatives, he chose to independently develop and commercialize its Ceredase/Cerezyme franchise. These were his first important commercial assets; they would never be partnered. Later, on the biotech boards he served, he would routinely recommend against partnering assets. To Termeer, independence equaled strength.

10. *Pay as You Go.* Like his father, Termeer was a Dutch mercantilist at heart—fiscally conservative and shrewd. He was not one to spend money he did not have. He sought to diminish Genzyme's dependence on fickle capital markets. And he did. He was not cheap, but you could discern his leanings when the conversation turned to money. The irony is that this thriftiness may have been at the root of the company's ultimate undoing—an unwanted takeover. When Allston failed, Genzyme had not invested sufficiently in a backup production source for its key products.

11. *Engage in the Moment.* Termeer had an enormous gift of bonding with most everyone he met, whether complete strangers or longtime colleagues. He was said to create a "personal contract" with executives, shop floor workers, board directors, and young and inexperienced new hires. His style was nonthreatening and fostered trust as it sought and recognized the best in people. After his inspir-

ing, uplifting, and optimistic encounters, it was often said he made you feel like you were "walking on air."

12. *Anything Is Possible.* Termeer was a leader who had big dreams and set high goals. His colleagues would identify the constraints in achieving them; he would exhort them to find the solution. He would stretch the rubber band to near the breaking point. He was not afraid to break conventions and smash preconceptions. Ever the optimist, Termeer saw life as being the glass half full, hiring great people, infecting his team members with positive energy, and establishing goals that unified and inspired them.

As Termeer was fond of saying, *"Fortunate breaks occur when you create an environment that has ample opportunity and the foresight to capture them."*

CHAPTER TEN

The Crisis

Standing by the floor-to-ceiling sliding window, arms crossed, Henri Termeer looked out over the Boston skyline from his capacious 12th-floor office. It was a sunny Tuesday afternoon, June 16, 2009. The news had hit the tape, its impact softened by a well-crafted press release that framed the day's news as a "temporary interruption." A viral infection of a bioreactor in its flagship cell culture facility in Allston had forced Genzyme to stop production. The plant needed to be disinfected. The FDA was informed and concurred with the company's decision.

From a distance, the concern seemed minimal. The company's share price hardly budged. Life went on, just another speed bump. Henri shrugged and said to analysts it would all be taken care of in 30 days. It would soon be back to full steam ahead. Disaster airbrushed.

Seasoned Wall Street biotech analysts, however, knew better. One former analyst and investment banker (now Chairman of Biogen), Stelios Papadopoulos, PhD, summarized their reaction, "We looked at each other and said, 'What is he saying? How on Earth can they rectify the problem, pass inspections, and be back up in manufacturing and supplying in 30 days?' It doesn't happen."

And for those down the ranks at Genzyme, they thought they fully well understood what lay ahead, even if Henri Termeer had other ideas. Forever the optimist, Henri had always been able to find a way through these disruptions.

Beyond the potentially troubling strategic and financial implications, the plant's temporary cessation in production was most ominous for the many patients whose lives depended on the medicines made there. This became even clearer on the heels of a company statement on July 22, some 36 days later. The R word—"rationing"—was introduced. Genzyme's medicines were their patient's lifelines, and each patient had come to rely on the company as a supplier they deeply trusted. Cerezyme, Fabrazyme, and Myozyme were all produced at Allston. And all treated rare diseases. And the two largest, Cerezyme and Fabrazyme, were single-source products.

The big irony of such a disruption at the "Allston Landing" facility, as it was called when it officially opened in 1995, was its proximity to the Harvard Business School, a mere mile up the Charles River. The building was divine, dubbed "the Cathedral on the Charles" by the rowers who passed by each morning. It was designed after a European church and put into service for the expressed purpose of making Cerezyme, a life-saving pharmaceutical. One would have thought Genzyme had long prepared for this scenario—a supply problem or even outage of a critical product. Whatever had happened to the concepts of safety stock and backup site redundancy? Had not some smart MBA sent word downriver?

Some patients were very upset, especially after the July announcement that certain drug supplies would be limited and require allocation. One particularly strident, disgruntled Gaucher patient activist hyperventilated, "People were hysterical at the time... . Everybody was pissed off because once again ... we were the last ones to find out and the first ones to die... . Their medical advisory committee came up with a pecking order of who would live and who would die basically, who got medicine, who didn't... . Fabry patients, they had no drug. I mean none, like two boxes... . Those people got screwed really badly. And the Fabry patients were even madder than we were. And we get our infusions together, so we knew all of them ... it was a difficult time and people got sick, yeah."

For Henri Termeer, this was his worst nightmare. Here was a leader whose core principles were individual responsibility, accountability, and placing the interests of patients first. His failure to secure a second source of supply for his patients was a monumental failing he would regret for the rest of his years. Over the ensuing 19 months, he took it hard, and very personally, as the crisis deepened. To his great credit, he never shirked the personal responsibility and accountability for the pain that was to follow. He knew he had let others down and would shoulder the blame. His health suffered, and Termeer's credibility would take a big hit.

But for now, the show had to go on. The company was facing a ghastly beast. In fact, at first, it was not even sure what beast it faced. But soon it became clear that the beast was instead a tiny, microscopic bug. The plant's contamination had likely been caused by a latent virus, Vesivirus 2117. It was found in the fetal calf serum Genzyme imported from New Zealand and used in its big tanks to feed the cells that expressed the company's enzymatic products.

After an intense review, it was concluded that a plant shutdown and decontamination was the only way forward. The reality of it all had hit the weekend of June 13. Mark Bamforth, Steve Kennedy, Blair Okita, Sandra Poole, and a coterie of Genzyme's other production leaders dropped everything and came together to create a plan for the site's remediation.

Poole's engagement was representative. She had even been at a Belgian amusement park, outside Brussels, celebrating her son's birthday when she had gotten the call to action. She recalled, "I got the call on Sunday from Steve Kennedy, and Monday morning, I was at the plant. For the next two months, we were in the plant pretty much seven days a week, sleeping barely a couple of hours a day. It was amazing."

They set about, an army of troops with an unlimited budget, to decontaminate the plant. All 180,000 square feet of it had to be dismantled. There was a frantic rush to clean it up. All available employees were deployed; their commitment was of epic proportions. The Doubletree Hotel next door became "home" for many. It was all hands on deck. This was about survival.

Vaporous hydrogen peroxide was used to sterilize the plant's six massive, multi-thousand liter stainless steel production tanks, and the walls in every hallway were bleached. Five miles of insulation, one mile of copper tubing, and 267 HEPA filters were replaced. Chromatography columns were reloaded by the dozens. Everything in the plant was viewed as a risk. Everything in the plant needed to be cleaned, sanitized, or replaced. Everything.

Remarkably, it took only eight weeks before the facility returned to service and the restoration of production capacity began. Termeer checked in with his Allston team each morning. His patients' trust had been rocked, and Genzyme was far from out of the woods.

In late summer, he had more bad news to communicate, but he tried to calm things as the plant reopened. His remarks that morning were directed at Genzyme's inventory supply shortage, "Reaching a decision on the work in process has been difficult for us as we balance the medical benefit of Cerezyme for patients with minimizing the risk to our newly cleaned Allston plant. In the end, we cannot take the risk of processing material that has any possibility of recontaminating the plant and setting back our ability to supply Cerezyme to patients. Now that we have resumed production, we are focused on the road to recovery."

The upshot of this decision was the destruction of over half of Genzyme's inventory, its safety stock, and its work-in-process pipeline. Sandra Poole, by then the newly installed head of the Allston plant, explained, "We lost a significant amount of a year's supply. We were already running at lower than normal inventory levels." But the discarded product had been deemed to be at risk. As a result, its Cerezyme and Fabrazyme supplies were nearly depleted. Total write-offs, exceeding $30 million, were eventually incurred, resulting in yet another hit to Genzyme's earnings.

Indeed, the usual marketing script had been flipped as Genzyme was forced to resort to "rationing meetings," a careful, tedious process of making sure none of the lives of its patients were lost. As one executive described the situation, the crisis had "precipitated the company from a honeymoon period into desperation."

These meetings were regularly attended by numerous Genzyme leaders, but their principles were guided by Termeer. Not only were those who took ownership of delivering the rare disease division's financial results involved, but also those responsible for the clinical, regulatory, and manufacturing/supply chain organizations. Termeer's order was to treat all patients equally, irrespective of whether Genzyme received payment for the product delivered or otherwise. The allocations were to be based on medical need, not financial value.

Caren Arnstein, Termeer's head of corporate communications, described the vibe, "Everyone was devastated. We all knew the instant we heard the news about the Allston plant. Without that plant, we couldn't supply patients. That is who we were and what we were all about. It was all about getting the patients the drugs they needed and here we were. We were their lifeline. There was this huge sense of 'oh my god, we've let down the patients.'"

Because of national differences in regulatory and other guidelines, different methods were developed by Genzyme for allocation in different territories.

For instance, in the United States, the allocation team would often evaluate individual patient's needs to determine the severity of their condition and the criticality in treating it. In the background, they would rely on a set of decision rules that favored the shipment of Cerezyme to children with more severe disease over older patients with less severe disease. Also, an emergency access program was set up for physicians to receive Cerezyme for patients who were in life-threatening situations. The Genzyme committee would make the final determination as to who would get what, subject to product availability.

In Europe, the regulators took a less forgiving stance. They insisted that its citizens get a full dosage, especially for Fabry patients whose symptoms often included impairment of vital cardiovascular, central nervous system, or renal functions. Dr. John Barranger, a NIH researcher in Roscoe Brady's lab at NIH who helped develop Fabrazyme, commented on the drug's administration, "Adequate dosing matters. If you don't take enough, you get sick."

This, of course, created its own uproar. Allen Black, a personal injury attorney who represented Fabry patients, had this to say in the *Pittsburgh Post Gazette*, "Europeans get a full dosage. Americans don't... . It's absolutely bizarre that Americans should essentially have no choice and be kept on a low dose... ." Black had a point. After all, the NIH had spent $4.1 million in U.S. federal funds to develop Fabrazyme.

Ed Kaye, Genzyme's Group VP for Clinical Development and a member of the rationing committee, would later explain his perspective on the patient community's reaction, "... people looked at Genzyme as a company with integrity and we were ethical, doing good science, trying to help patients. People remained loyal to Genzyme."

Geoff McDonough, MD, one of the executives responsible for Genzyme's rare disease franchise, tried to reassure the patient community in a prepared statement, "These actions are intended to preserve inventory for the most vulnerable patients and to ensure global equity in this extremely challenging time for patients and physicians."

The company had originally communicated that the supply shortfall would end in October. By August, however, it became clear that this timeline was not going to be met. The company was forced to amend its projection; new product would not be released until later in the year, possibly not until the end of December, adding two more months of uncertainty.

Later, it would come to light that Genzyme had been facing supply pressures for some time. Undetected, the insidious Vesivirus had been attacking Genzyme's cell cultures for months, impairing its cell growth and production yields.

Patients read the news; they were desperate. Jack Johnson, Executive Director of the Fabry Support and Information Group, or FSIG, remembered the situation well, "That was a very difficult time. There were patients suffering a great deal around the country as well as new patients overseas. As a patient advocate, we wanted to know what was going on and receive up-to-date information as soon as we could possibly get it. Unfortunately, what FSIG and others would get was stale, recycled information after the company's investors had been briefed." Johnson conceded it had not been Genzyme's fault. "That's the way the system works," he said. But it stung nonetheless.

As was customary in these situations, regulators had also remained vigilant as Genzyme was rebooting Allston Landing. To monitor its progress, the FDA field office in Stoneham, Massachusetts set up an Allston site inspection, which commenced on October 8. This was part and parcel of the enforcement of the FDA's Good Manufacturing Practices (GMP), its stringent code for ensuring drug safety and quality.

Upon its completion five weeks later, FDA representatives visited Allston to read out their conclusory findings. A Form 483 Report was issued on that dreary, rainy Friday, the 13th of November. They had asked Henri to be present for the meeting. In its 22 pages, there were 49 infractions cited, an extraordinary number. Termeer sat expressionless while Thomas Arista, the lead Field Investigator, read the violations one by one, a recitation that took more than a half hour. As Sandra Poole recounted, "This was a dark, dark moment for Henri." Reality was setting in. A few months later, the FDA notified Genzyme that additional Enforcement Actions would be taken.

A fundamental question that often came up as Genzyme's community of patients and supporters pondered their plight was why the company had not built a backup manufacturing facility. It seemed unfathomable.

One answer was obvious: the cost, a half a billion dollars. But since Allston's opening 14 years prior, Genzyme had been growing exponentially, both organically and inorganically, and its profits were soaring. The company surely had the financial resources to build it.

Since 1995, the company had made numerous updates and improvements to the facility, including the construction of a large new wing to upsize its capacity. Nonetheless, although long planned, the sizable investment in a backup site to produce Cerezyme and Fabrazyme was often bumped from the top of the capital expenditures list.

One prominent reason for deferring the investment was attributable to nine infants and a clutch of genetically modified rabbits, all part of an experimental clinical trial being held in Europe in the early to mid-2000s. The infants had Pompe disease, and the rabbits produced milk that contained a raw enzyme that, once removed from the milk and purified, enabled the babies to be treated. Based on the study results, the enzyme appeared to essentially eliminate their risk of death.

Learning of this potential medical breakthrough, the parents of other Pompe patients came forward, demanding treatment for their children. In some countries, they even went on television and chained themselves to health ministries to publicize their cause. They demanded that the company produce enough enzyme to treat their children and all the others who needed it, not just the first nine.

To address these concerns, a shift in Genzyme's production and supply strategy was made, and the build-out of a backup site in Framingham, Massachusetts was abruptly put on hold. This enabled the reallocation of funds for the construction of Pompe-related production capacity in Geel, Belgium at a site Genzyme had acquired from Pharming Group NV in December 2001. And in the meantime, the rabbits, which were fed and supported at this site, could continue to produce drug product for the nine kids.

Converting this small Belgian plant for the large-scale production of what was to later become Myozyme would take an enormous investment, and Termeer prioritized it because he had seen the drug's dramatic effectiveness in treating Pompe patients. One Genzyme executive later remembered Termeer's assertion that hundreds of babies' lives had been saved as a direct result of establishing this priority.

But in hindsight, this was also a consequential, fateful decision that, although saving many lives, contributed to the inadequate backup of the Allston plant and the company's eventual inability to provide a steady, secure supply of drug product to Gaucher and Fabry patients. Its repercussions would not be fully known for seven years, a stretch of time during which the question of building an Allston backup site was always in play but never sufficiently addressed.

In looking back, it was obvious that additional backup production capacity—in Framingham or elsewhere—should have been built to ensure the company's ability to supply Cerezyme and Fabrazyme. The fact that the investment in a backup site was continuously deferred during the 2001–2009 period revealed, at the very least, a level of parsimony gone too far. At worst, it reflected an uncharacteristic blind spot due perhaps to corporate hubris or incomprehension of the disaster that would ensue in the event of a supply outage. Years later, one executive rationalized that it was nothing more than an expression of overconfidence in their capabilities and the fact that "we had never failed before." Sadly, they now had, and the costs of this mistake were tragic and far-reaching.

In parallel with the Allston production crisis, Termeer began facing a base of restive Genzyme shareholders. Its shares were languishing. Sales and profit targets were being missed because product was not being shipped. Financial guidance was being lowered. How was Genzyme going to make its numbers? It was now a $4.6 billion corporation and among the Fortune 500, but could this be sustained? In late July 2009, Genzyme adjusted its range of financial guidance for the year ending December 31, 2009. It was revised to $4.6–5.0 billion, down from $5.25–5.35 billion. The company ultimately delivered $4.5 billion for the year. The revenue gap resulted primarily from the triad of products that had been halted at Allston—Cerezyme, Fabrazyme, and Myozyme. Each missed its target.

Looking back, these shortfalls revealed a basic flaw in the strategy that had been foundational to Termeer's steady, opportunistic acquisition of businesses over his 28 years in leading Genzyme. By the time Genzyme was sold in 2011, the company had done more than 30 acquisitions—far more than any other biotechnology enterprise of its generation—in a wide range of fields including renal, multiple sclerosis, hematology/oncology, biosurgery, pharma intermediates, and genetic testing.

From the beginning, analysts had postulated that Termeer's true strategy centered not on rare disease, but on diversification. His strategy was to construct a puzzle of disparate pieces that somehow would fit together under the heading of "Patient Care." These various pieces would insulate

Genzyme from the cyclical vagaries of markets, the disruption of competitive new product launches, patent expiries, and operational hiccups like Allston. One analyst went so far as to posit that Henri was trying to create over decades another J&J—a highly diversified international healthcare products company.

What became evident over the years, however, was that the investments that seemed to be working best, for Genzyme and its shareholders, were those made in treating rare diseases. It is not that the others, especially Renagel, were not able to contribute to the top and bottom line. Many were, just not to the same degree.

So although the nonorphan businesses had provided some ballast as the Allston plant disaster was being stabilized, they were hardly the value-creating engines that Cerezyme, Fabrazyme, and the other rare disease therapies offered.

The analysts and activist investors sharpened their pencils watching the company's fortunes stagger. The company's downward spiral exposed Genzyme's underbelly. There were underperforming units in the portfolio. Before long, the pressure would build to divest them, something Termeer had assiduously resisted over the years as it would impair his bigger dream for Genzyme.

This strategic weakness invited investors' criticism and activism. Ralph Whitworth of Relational Investors was one. Whitworth had come onto the Genzyme scene in 2009 as news hit the papers of not only its Allston troubles, but also as awareness of the underappreciated value of certain of its various business units became recognized. The company's shares were drifting lower that fall as the FDA setbacks and missed financial guidance were closely watched by investors.

For the activists, this had the makings of a breakup candidate. Whitworth had already made a fortune in breaking up companies. He actively sought companies deemed to be underperforming because of poor capital allocation discipline. Boone Pickens had been his mentor. His scalps included the CEOs of Home Depot and IBM who had been forced out at his behest. Although reputed to be a "quieter" activist, Whitworth carried a big stick.

Relational Investors began accumulating Genzyme's shares in late 2008, and by the end of 2009, it owned approximately 4% of the corporation, worth about $540 million at the time.

Whitworth had been discussing Genzyme's fortunes with Termeer over the course of the previous summer. To say they were "friends" would have been a stretch. "Semi-adversarial" was perhaps a more accurate descriptor, used by one Genzyme insider who had observed their relationship's dynamic. But a dialogue had been established between the two as one would expect of a CEO and a major shareholder.

As the Allston site reopened and the FDA delivered its 483 citations, Whitworth commenced a months-long escalation of pressure that would catalyze far-reaching changes in the composition of Genzyme's board and management team, as well as the company's management practices. Since the mid-1980s, Henri Termeer had had a free rein in leading Genzyme, shaping and guiding its strategic direction, its decision-making apparatus, its human capital practices, and its public persona. That was all nearing an end as external forces began to close in around him.

In many ways, the contest harkened back to the games Termeer had played in the chess halls of Tilburg, those that had honed his ability to look around the corner, strategize, forecast, and prepare. But this time there was a difference as the king on his chess board had landed in a position weaker than any he had encountered since the mid-1980s.

On December 10, 2009, as news on a host of fronts needed updating, Termeer issued a letter to his shareholders, informing them of the many developments at Genzyme. In classic Termeer fashion, it forcefully and optimistically presented his case that Genzyme would "emerge a stronger company that is better prepared to deliver on our commitment to sustainable growth."

The letter replayed the year for its readers, factually stating the conditions that had been successfully navigated. It included not only an update on the Allston plant's turnaround and its outlook, but also the changes in Genzyme's operational leadership, including four new senior managers brought in from outside the company to run various key functions. Termeer also, for the first time, gave readers an insight to his own succession plan: David Meeker, John Butler, and Mark Enyedy, all Genzyme insiders, would share oversight for the company's commercial and manufacturing operations. Of the three, Meeker got the largest share of the responsibility: EVP of Genetic Diseases, Biosurgery, and Corporate Operations, which included Allston.

Tucked into the letter was also the appointment the day before of Robert Bertolini to Genzyme's Board of Directors. Bertolini had been the CFO at Schering-Plough at the time of its $41 billion sale to Merck & Co. The deal had been a huge success for Schering shareholders, and it had closed just one month earlier. Bob was one of its principal architects, and he was viewed as an investor-friendly director with good judgment. Little fanfare was brought to the appointment, but it was a signal of Termeer's openness to amending his board of directors and providing investors with another independent voice at the table.

Days later, Ralph Whitworth had taken it all in and decided to make his move. On December 15, he rang up Termeer and formally requested a seat on Genzyme's board. It was the holiday season, and the two spent much of it discussing Whitworth's request.

On January 7, the week before the year's most important life sciences investor conference, the 28th Annual J.P. Morgan Healthcare Conference, Genzyme announced that it had entered into a "mutual cooperation agreement" with Relational Investors. Subject to certain conditions, including a standstill provision, Whitworth had agreed to defer his appointment to join the Genzyme Board until later in the year, November. He also agreed to support those who would be nominated by Genzyme to stand for election at the company's upcoming annual stockholders' meeting in June.

On the following Monday, January 11, the opening day of the J.P. Morgan conference, Bob Carpenter was re-elected as Genzyme's Lead Independent Director, a role that had been strengthened and formalized the prior week. The Board had voted to expand Carpenter's role to "include responsibilities that are similar to those typically performed by a chairman who is not a company CEO." It was expected that Carpenter would serve in this role for three years.

The following day at 9 a.m. PST, Termeer presented Genzyme's 2009 performance and its 2010 outlook to a standing-room-only crowd in the Grand Ballroom at the St. Francis Hotel. Summarizing the past year's developments, the challenges ahead, and the shortfalls in Genzyme's financial results could not have been much fun. But Termeer, ever the optimist, found a way through the San Franciscan fog, "In 2009, we continued to successfully execute across our diversified businesses," he said. In the accompanying press release, of the eight annual highlights cited on its first page, not one spoke to Cerezyme or Fabrazyme. These were buried in the text that followed.

It was not Termeer's best year at J.P. Morgan, but he was by now a member of the royal family—of biotech, that is—and recognized as its longest tenured and one of its most successful CEOs. As a gifted communicator, he pulled off his half-hour presentation with the humor, aplomb, and the charisma all knew of and expected from him.

Meanwhile, Genzyme's shares had closed on January 5, 2010 at $48.24. By the close of trading on January 12 they were trading at $52.98, representing a one-week gain of nearly 10%.

Over the next month, things quieted as the company set about closing its year-end books and preparing for the Annual Meeting and Proxy season. Behind the scenes, however, investor activist Carl Icahn had been accumulating Genzyme's stock. He had begun building his sizable position in mid-August 2009. If Ralph Whitworth was the quiet activist, Carl Icahn could be expected to take up the slack and make enough noise for the two of them.

Icahn was not one to disappoint. On the morning of Monday, February 22, Icahn Partners notified Genzyme that it intended to nominate a slate of four individuals for Genzyme's board of directors at the company's 2010 stockholders' meeting, which at this point was scheduled for May 20. Genzyme was already well along in making its preparations. The company's Initial Preliminary Proxy Statement was filed with the SEC on March 11, and it became clear that a proxy fight with Icahn Partners loomed. The annual meeting could turn into something just shy of a riot.

Icahn filed his own proxy statement on March 23, nominating his slate of four Director nominees. It confirmed that Icahn Partners and its affiliates had acquired nearly 4% of Genzyme, worth about $575 million on the open market. Publicly, Termeer responded by saying about the only thing he could, "... we are open and responsive to shareholder input, and we welcome a constructive dialogue with Mr. Icahn."

Somehow Termeer kept his cool among the constant winds that were buffeting Genzyme. The pressure was intense and building, and it would get worse, but Termeer, being the strong leader and gentleman he was, handled himself and his adversaries with toughness, but also with a measure of equanimity and grace. His team and his board remained firmly behind him, even if the activists and regulators were launching grenades on a seemingly daily basis.

On Wednesday, March 24, 2010, Genzyme and Termeer would hit their low water mark, perhaps of the entire saga. The FDA had called again, this time with some news.

A Genzyme press release later that day would explain that a Consent Decree of Permanent Injunction, a civil enforcement action resulting from its Allston failures, was being prepared. Termeer and two of his associates would be Named Defendants. The details of the filing in the District Court of Massachusetts were due to be released in several weeks.

CHAPTER ELEVEN

Changing of the Guard

It was early on March 24, 2010, and Allston's top quality executive Blair Okita, PhD, was on the way back to his office. On departing his morning meeting at Genzyme Center, he bumped into a member of Henri's inner circle and Executive Committee member. "I need to talk with you," the executive said solemnly. They ducked into a quiet office.

The day's news was not good. The FDA had informed Genzyme that it would be taking enforcement action against the company for its continued violation of the FDA's GMP code. It was a ringing rejection of Genzyme and as clear a signal as could be given that the agency had lost confidence in the company's ability to solve its manufacturing and quality control problems.

Termeer and his team hastily issued a news release and convened an 11 a.m. conference call to update Genzyme's investors. The company announced that the FDA "intends to take enforcement action to ensure that products manufactured at the plant are made in compliance with good manufacturing practice regulations. The FDA enforcement action will likely result in a consent decree," meaning it would result in the agency's oversight of the plant and stiff financial penalties.

Six days later, David Meeker was named COO of Genzyme, a position that would focus on leading the company's commercial organization, country management, and market access functions. Given the stakes, Termeer reshuffled the reporting lines, opting to place himself closer to the production and quality organizations. Two recently hired production and quality leaders, Scott Canute from Eli Lilly and Ron Branning from Biogen, would now report directly to Termeer.

On the investor side, the news fallout served to heighten the pressure already being wrought upon Termeer and his leadership team. It is hard to imagine how Termeer pressed onward. Between patients' discontentment,

FDA actions, unhappy investors, and a brewing proxy fight, the temperature had to have been very high.

Icahn Partners was amending its own Preliminary Proxy Statement, preparing for battle. And Ralph Whitworth, not willing to subordinate his interests to Icahn, was now rattling his saber; he wanted to renegotiate his cooperation agreement and accelerate his appointment to the Genzyme Board. Genzyme concluded that it needed more time to prepare for its annual meeting and delayed the scheduled meeting four weeks to June 16.

Not a week had passed before Genzyme capitulated, agreeing to appoint Whitworth with immediate effect and thereby increasing its board from nine to ten members. Whitworth would continue to agree to a standstill. He would later be joined on the Board by another Director whom he would choose. It was agreed that his designee, whose appointment would take place at the upcoming annual meeting, would have "expertise in pharmaceutical or biologics manufacturing or quality control operations... ."

At this point in the drama that was unfolding, even a leader of Termeer's extraordinary capacity and optimism was mightily challenged. And there was little relief in sight. Those around him attested to his strength. He did not wear his concerns on his sleeve.

To them, he remained "Henri." But two weeks ago, it had been the Consent Decree; this week it was Whitworth; next week it would be the release of Genzyme's disappointing first quarter results.

The numbers would show another down quarter when compared to the same prior year period. Sales forecasts were largely not being met; profit margins were under pressure. And the preliminary draft terms of the Consent Decree were released. An up-front disgorgement payment of $175 million was on the table, with other penalties threatened for noncompliance.

The $175 million penalty was confirmed on May 24. Genzyme, Termeer, Okita, and Branning were all Named Defendants in the enforcement action. The FDA had wanted to charge other Genzyme executives, but the negotiations resulted in their removal from the civil suit. The ammunition that this action gave to Icahn and others who were advocating change in the company's leadership was unmistakable.

Two days later, Icahn disseminated to Genzyme's stockholders a 62-page PowerPoint presentation. It was his case for removing the Genzyme Board and evaluating its executive team. It condemned the company's performance under current management. It was his version of a manifesto for change. It was entitled "Genzyme, Time to Change the Old Guard." Days later, it was followed by a letter Icahn penned to Genzyme's shareholders, seeking their vote in the proxy fight that had broken out into a barroom brawl.

Genzyme was quick to respond, having its own PowerPoint perspective to share. It provided a considerably more upbeat outlook, criticized the methodology in Icahn's analysis, and mentioned the company's plans to divest three of its noncore business units—genetic testing, diagnostics, and pharmaceutical intermediates.

By now, both parties had dug in and the gloves were off.

The media took notice, and Genzyme enjoyed substantial support for its intransigence in placating Icahn. Howard Anderson penned an op-ed in *The Boston Globe,* castigating Icahn, "Carl Icahn's Battle to Take Down Genzyme." The essay cited greenmail, opportunism, and "Icahn's hunger for short-term profits." Earlier, on April 13, *The Globe* had run a story headlined "Keeping the Life in Life Sciences." Its authors defended the 400 biotech companies, including Genzyme, that were clustered in Massachusetts, employing 77,000 people and contributing notably to the state's economy.

In a NYT blog, sponsored by *The New York Times,* Harvard Business School professor and former Medtronic CEO, Bill George, cited Icahn for seeking to maximize "short-term value in an industry where enormous investments and extended time frames are required to create long-term shareholder value." It also printed a letter from the Executive Director of the Hide & Seek Foundation for Lysosomal Disease Research, Jonathan Jacoby, who expressed his appreciation "to everyone at Genzyme for dedication to finding cures for lysosomal disorders."

Maybe their support had turned the tide in Genzyme's favor as Icahn and the company declared six days later that they had found a solution. Whatever the case, Icahn agreed to seat only two, instead of four, new Directors on Genzyme's board, and Icahn Partners would immediately cease all efforts related to their proxy solicitation.

The annual meeting was convened on June 16, and the board was upsized to 13 sitting Directors. Dennis Fenton, Steven Burakoff, and Eric Ende joined the board, each of whom had been duly nominated and designated by the activist investors. Including Ralph Whitworth, there were now four designated activist investor-Directors seated alongside the nine legacy Genzyme Directors.

In the background, Chris Viehbacher, CEO of the French pharmaceutical giant, Sanofi, and his team had been evaluating acquisition targets in the United States to advance the company's strategic plan. At the time, Sanofi was primarily "a small molecule company," keen to diversify its risk away from the massive patent cliff that faced some of its largest, most profitable brands, such as Plavix, Lovenox, and Lantus. The company was also looking for a way into the U.S. biotechnology business and, more specifically, into Boston, the emerging center for worldwide biotech R&D. In the words of Chris Viehbacher, "the ideal transaction would be accretive to earnings and involve the acquisition of a larger company." Viehbacher and Henri Termeer had met previously at various PhRMA Board meetings, and although they knew each other, they did not know each other well.

After introducing the prospect of Genzyme as one of Sanofi's ideal targets, the Sanofi Board requested that Viehbacher reach out to Termeer. On May 23, in the midst of the Icahn proxy fight, he called Termeer to offer his support and to open a dialogue about Sanofi's interest in a potential transaction. Termeer demurred but suggested that a structure akin to what Hoffmann-La Roche and Genentech had earlier designed in their initial tie-up could be of interest. He was, however, in the middle of a proxy fight, and the timing was not right. He suggested they continue the conversation sometime after the annual meeting. Viehbacher took note and signed off.

A month later, at a regularly scheduled board meeting, he briefed the Sanofi Directors on his conversation with Termeer. They agreed that Viehbacher should reapproach Termeer to explore things further. By now, the activists were installed on Termeer's board and asserting themselves, the remedial demands of the Consent Decree were being implemented, and patient advocates remained vocal in their unhappiness with Genzyme, even after Allston had rebooted and production had resumed.

To this last point, there was a critical shareholder communication that Genzyme had issued at the end of June that played into Sanofi's hands. Genzyme posted an update on the product supply website it had created for patients and physicians who remained concerned about the adequacy of Genzyme's production recovery. The posting, which occurred June 29, 2010, indicated that the acceleration in the plant's production capabilities had essentially stalled. Supplies for Cerezyme and Fabrazyme were not expected to grow in July, and in the case of Fabrazyme, no growth in inventory was forecasted for the next three months. This unsatisfying outlook fed into the activists' case, as well as Viehbacher's.

This pause in Genzyme's manufacturing recovery may have provided the opening needed to encourage Sanofi to move more aggressively. It was an expectations problem. Genzyme had set expectations that could not be met. The plant recovery would take months longer than had been forecasted.

Termeer and Viehbacher jockeyed for position as the summer wore on. Viehbacher would try repeatedly to get Termeer to engage with him. Termeer avoided contact where he could. Viehbacher publicly announced a preliminary offer he had written to Termeer on July 29, valuing the business at $69 per share. Sanofi's offer was flatly rejected by Genzyme. Termeer and Viehbacher met on August 20 to discuss the offer. Their respective M&A advisors met on August 24, feeling each other out on process steps. The Sanofi bankers were keen to commence due diligence. The Genzyme bankers hid behind their client; they said they were unauthorized to permit it.

If there was one thing Termeer had always cherished about Genzyme, it was the company's independence. One of his favorite sayings was, "Independence equals strength." If Genzyme were to lose its independence, it would not be without a fight. Sanofi sent Genzyme a second all-cash written offer at the same price on August 30, complaining about Genzyme's unwillingness to engage. Within 24 hours, Termeer shot back a letter rejecting Sanofi's offer as inadequate.

There were several volleys of meetings and phone calls during September. As one of Termeer's top executives later confided, "Henri didn't want to sell the company. We thought we could basically outlast them."

But by early October, after consulting his board, Viehbacher decided he could wait no longer. He needed a signature deal in the U.S. biotech market that would address Sanofi's strategic imperatives. Genzyme met his criteria and had become Sanofi's No. 1 target. He kept hammering away. He called Henri on the morning of October 4. Sanofi was launching an unwanted, all-cash takeover bid for Genzyme at $69 per share. The tender offer would go out at 9 a.m. EDT before the U.S. equity markets opened.

By now, Genzyme had instructed their bankers to determine if there were alternative buyers for the company. It was decided that a formal auction would not be conducted, but Goldman Sachs and Credit Suisse were directed to discretely identify a buyer who could step up and offer a topping bid.

At the time, the list of potential bidders who could write 25 billion dollar checks was limited, and several of the likely prospective bidders had recently transacted. They were not in a position to make or digest another major acquisition. Other prospective bidders were still not convinced that a business model that relied on charging patients annual six-digit sums for drug therapy was a model that fit their own. And still others were sufficiently concerned about the uncertainty surrounding Genzyme's manufacturing challenges and the Consent Decree. They could not get comfortable with the risks.

Other factors weighed on the company's outlook, among them were human capital issues and sustaining employee morale. Genzyme's financial results for the third quarter of 2010 were announced in late October. Revenues were up slightly over the comparable prior-year period; the genetics testing and diagnostic businesses that had been put up for sale were expected to be divested by year end. To boost the company's profitability, operating costs were being heavily scrutinized. The activists wanted the sails trimmed and personnel costs cut.

Since its founding in 1981, there had never been a layoff of employees at Genzyme. So as the company implemented the first phase of a "workforce reduction plan" in early November of 2010, its employees were aghast. The memo announced that the target was to eliminate 1,000 positions by the end of 2011. The first phase would address 392 positions. It was part of a larger strategy to "increase shareholder value." The combination of activist pressure and Sanofi's pursuit was forcing Genzyme to take action. Morale flagged.

Based on the reaction at Genzyme Center, one would have thought the world had come to an end. The bars along Broad Canal next door to Genzyme Center filled by 3 o'clock that afternoon with workers reeling from the news.

Tomye Tierney had been one of those asked to clean out her desk. Henri sent word and had asked to see her. It was the 20th anniversary date of her employment at Genzyme, Friday, November 5, 1990. She reminisced, "Henri had never laid anyone off ... they took so many long timers. That's where they could take the biggest savings. Henri had me come up to talk to him right away, and he was so good. I can't remember exactly what he said ... something like, 'Believe me this is the best thing for you. It's just going to get worse... .'"

Henri cared deeply about his colleagues. This was a body blow that pained the Genzyme community. If there had remained any doubt, it was now confirmed. He had lost control of the company.

In early December, Genzyme announced the sale of its genetic testing business to LabCorp for $925 million in cash. It was a divestiture that had been conducted to placate the activist-Directors. Without pressure, he would never have divested the business. The diagnostics and pharma intermediates businesses were next up. They would likely be gone in six months.

Approaching the end of the year, Viehbacher and Termeer remained in conversation; Sanofi continued its tender offer which eventually was extended into 2011. His closest advisors, especially Peter Wirth and Zoltan Csimma, Genzyme's CHRO, were among those from whom he regularly sought advice. He kept his cards very close. He shared little. Alone at times, he would pace in his Marblehead study at night, deep in thought.

The New Year dawned. It was J.P. Morgan again; its 29th Annual Healthcare Conference would open its doors in a few days. Termeer had had the holiday season to ponder Genzyme's future. A white knight bidder had not emerged. Its business, although in recovery, was still lagging behind its targets. Termeer was approaching his 65th birthday.

Although he had never planned to step back, much less retire, Termeer was edging toward a decision. Sanofi, the regulators, the activists—they had worn him down.

As he evaluated his options, he was faced with a chessboard that he did not like. He could auction the company, but in their own way, Genzyme and its advisors had already tested those waters. If they ran it out and Sanofi walked, Genzyme would be in a vulnerable position. Their stock price might collapse, their investors' expectations would not be met, and Sanofi could swoop in, acquiring the company at a much lower level.

So Termeer decided that under those circumstances, he was not willing to take the risk. He could not. He never explained his thinking explicitly, but were he to continue his quest for independence, he was in the position of having to ask his Board and his investors to "stay with me." But he could not make the ask, especially if he were, maybe, going to step back, let alone retire. So, as explained by one of his closest advisors, "He had run out of options. He took the bull by the horns and negotiated the deal with Viehbacher."

That year, Termeer's presentation at J.P. Morgan was professional but little more than perfunctory. He was saving his dry powder for another day.

By now, Termeer had confronted reality like few others at Genzyme. To his benefit, of course, he held all the cards and could see the endgame was near. As he entered 2011, the conversation within himself had led to his reconciliation with the inevitable. He had seen it coming for several months. As clear as his options had been in 1983 when he took leave of Baxter for Genzyme, so too were these he faced 28 years later.

Through the pain and tumult of the last 19 months, he had somehow made peace with what had occurred, and he now, reluctantly, was turning to his future. Although difficult to accept, he had come to grips with this fateful outcome. He would move forward in a manner that neither he nor those around him could have imagined a year earlier.

The World Economic Forum's annual four-day extravaganza at Davos, Switzerland had begun. It was the last week of January. Private jets were parked three-deep at the local air fields; 1,700 jets would drop off their passengers. A fifth of the grandees would helicopter in from Zürich International Airport or other nearby helipads. Leaders from every corner of the world were in attendance. A who's who of prime ministers, presidents, and captains of industry attended. Warren Buffett, Bill Gates, Jeff Bezos —they were all there. If you led an important business, it was likely you were there. George Soros had opened the Forum with a stern warning about the U.K. economy. Tim Geithner, the U.S. Treasury chief, would be meeting with British Chancellor of the Exchequer, George Osborne. Russian President Dmitry Medvedev was cutting his visit short, flying back to Moscow in the wake of a terrorist bombing.

Henri Termeer and Chris Viehbacher had also made the trek to this hamlet tucked away in the snowy Swiss Alps. The odd part is that neither initially knew for sure that the other would be there, although they likely suspected as much. As fate would intervene, that year's Davos Conference would turn out, for them, to be more than just a series of global policy discussions. There was business to be done.

Victor Dzau, then Chancellor for Health Affairs at Duke University and one of Chris's old friends from Durham, was at the time one of Genzyme's longest serving Board Directors. He became aware of the two CEOs' presence at Davos and saw the opportunity to set the wheels in motion. He suggested to Henri that they have dinner; he would call Chris to propose their evening together. To lighten the tension, he would join them, as would Sanofi's CFO, Jérôme Contamine. Henri concurred, the contact was made.

The four of them met at 6:30 p.m. outside the town congress hall, Davos's Kongresszentrum. It was a short but steep walk up Buolstrasse to the Waldhotel, an elegant, family-owned establishment with a gourmet restaurant overlooking the town below. Dr. Dzau later laughed, "None of us would admit that we were sweating bullets (from climbing the hill) ... and this was winter." Henri and Chris walked together, arm in arm, setting the tone for the evening's discussion.

They sat down for dinner in a private room at the back of the restaurant. Light snow was falling. It was a cold midwinter night, 20°F outside.

Accompanied by Dzau and Contamine, this was the night it would all come together. Henri Termeer and Chris Viehbacher had reached the point where the four corners of a deal had, already, nearly been determined. A process that had started eight months ago had run its course. There were details to be discussed and agreed, but they had each concluded that their companies should be combined.

They had grown to respect each other. In the words of Peter Wirth, "Chris played a tough hand, but he played it honorably. He knew what his position was... . Henri wanted the acquisition to be successful because he cared a lot about Genzyme."

Termeer had come to believe that Viehbacher really understood what Genzyme was all about. He could be trusted to continue its legacy and build on its good name, putting patients first. He would do his best to protect the values that Termeer had tirelessly built over nearly 30 years. Over the last six years of Henri's life, Viehbacher and Termeer would become good friends. They socialized with each other, got to know their respective families, and enjoyed each other's company. A bond developed as did an ease of friendship.

The most important of the deal terms were decided over dinner and a bottle of Château Pétrus. It was now up to the bankers and the lawyers to iron out the remaining points and package the accord. An all-cash offer of $74 per share with Contingent Value Rights (CVRs), kickers for certain successful postclosing events, would be made. This valued Genzyme at $20.1

billion, excluding the five CVRs, which might add another $3 billion to the purchase price. To celebrate this important occasion, Viehbacher would later sign and give Termeer a bottle of Château Pétrus that resides in the Termeer wine cellar to this day, a memento for eternity.

The two left the dinner table in a reserved but friendly mood. It had been a conversation to clear the air, reach a deal, and set the future, a "new beginning" as they would cast it a few weeks later.

They walked outside the Waldhotel into the fresh Alpine air. In the slick conditions, and in the manner in which they entered, arm in arm, they navigated their way down the hill, sometimes walking backwards to avoid a fall, with Dzau trailing but ready to catch them if they lost their footing. Although one story of hope was ending, another was beginning. The torch would be passed. Genius, power, and magic had each played their part in the bold drama that had unfolded.

For all of the sadness and disappointment that had accompanied Termeer's final two years and his ultimate relinquishment of Genzyme to Sanofi, there would be vitally important, life-defining opportunities to greet him on the other side. These would derive from Genzyme's overwhelming success, Termeer's leadership and immense stature, his generosity, and the catalytic event that was Sanofi's acquisition of Genzyme. This event would unlock waves of energy, talent, innovation, and capital formation rarely witnessed in the history of the life sciences. In ways that even Termeer, the visionary, could not have imagined, they would enhance his already formidable legacy.

On February 16, 2011, in a joint press conference, Henri Termeer and Chris Viehbacher announced that the companies had reached a definitive agreement, approved by both Boards. Sanofi would acquire Genzyme for $74 per share plus Contingent Value Rights. It was indeed to be a "New Beginning."

CHAPTER TWELVE

The Diaspora

On April 8, 2011, Genzyme's shares ceased trading on the NASDAQ stock market. The Sanofi merger had closed. More than $20 billion in purchase price consideration was distributed to Genzyme's shareholders, including its employees. Three billion more waited in the balance if Genzyme could meet certain postclosing performance criteria. Bloomberg reported that Termeer's payout at closing had been $158 million.

It would have been easy for Termeer to conclude that his day at the office was done. He could retire, travel with Belinda, serve on boards, and spend time with his children. Instead, however, he chose a different course. He was not ready for retirement, far from it. There were scientific discoveries, new genetic diseases identified, new orphan drugs, and new companies—and it seems the people involved in most of them wanted his help.

Genzyme's corporate independence had been surrendered. As Gustav Christensen, a former Baxter contemporary and colleague, would describe a conversation with Termeer, "It was not exactly how he wanted it to end." But months before, Termeer had faced the possibility of this denouement, and when it finally came, he was ready to accept it with his characteristic optimism. David Meeker described Termeer's outlook, saying, "He exited as he should have, the senior statesman.... It was clearly rejuvenating for him."

Gail Maderis, a Termeer protégé and currently President & CEO of South San Francisco–based Antiva Biosciences, was leading BayBio at the time of the merger. It was the organization that supported the Bay Area's biotech community, and she was plugged into the entire scene. In 2004, for those Genzymers attending J.P. Morgan's 22nd Annual Healthcare Conference, she had inaugurated what would become an annual gathering of its alumni and active executives to catch up and kick off the New

Year. With Sanofi's acquisition of Genzyme, the 2012 30th "J.P. Morgan," the conference's eponym, was going to be a new experience. And many thought Maderis's party might be uncomfortable after the acquisition. Would it sunset?

Thousands of biotech mavens—CEOs, CFOs, BD&L experts, financial analysts, scientists, bankers, recruiters, lawyers, and reporters—would attend J.P. Morgan each January. This premier conference was the granddaddy of them all, having been founded in 1983 by one of the Four Horsemen of biotech investment banking, Hambrecht & Quist, a specialty firm that later became a 1999 Chase Manhattan acquiree. (In the following year, Chase Manhattan Corporation acquired J.P. Morgan & Company, later rebranding the corporate enterprise, JPMorgan Chase & Company.)

The conference's original raison d'être was to connect institutional investors with the firm's biotech clients, a mission that long ago had been eclipsed because of its burgeoning size and the professional diversity of those whose attendance it attracted. It was now a happening, and every day was an extravagance of biotech engorgement.

In that particular spring of 2011, Maderis was pondering whether or not to continue her annual Genzyme gabfest. Henri had attended every one since 2004. Maderis said, "Henri was always the first guest there and the last to leave."

Her inaugural event had been held at a rented apartment. By now, she had moved to an elegant home on Telegraph Hill, and the crowd would be 100 in number, but the tone never changed much. "Henri was very Dutch in the way that he managed expenses. Genzyme always funded the event. I would go to Costco and buy the wine and beer and keep the expenses to a minimum."

As the planning for next January's event began, she had to confront the fact that Genzyme would be owned by Sanofi. And with the handover in ownership, Sanofi executives would want to attend, especially if they were footing the bill. She pondered if it was not time to let it all go, thinking, "perhaps someone else should host it."

Maderis, at the urging of Jim Geraghty, however, concluded that in staying close to Genzyme's true values, the event must continue and it was her responsibility, "We really were a very close-knit group, more like family than just colleagues. We had to move forward as we had before."

She wondered how it would work with Henri and Chris Viehbacher both in attendance. "Awkward," she thought.

"Well I have a roof deck. I have my living room. So I invited all the Sanofi executive team... . I can leave Chris in the living room and put Henri on the roof, and everyone can talk and mingle with a little bit of distance."

In hindsight, the party would go off without a hitch. In reflecting, she continued, "I think it was one of the most important alumni parties we ever had. Chris and Henri are both just tremendous gentlemen and diplomats. They set a very congenial tone.

"And I think Chris saw a very different side of Henri at that party. He realized two things. One was the power of a vast web of connections, the Genzyme 'Diaspora,'" as *Boston Globe* reporter, Robert Weisman, would later describe it in a July 2015 article. The other was the Genzyme culture in action. "We were more than just work colleagues. There was a camaraderie that developed at Genzyme and a culture that was just so strong. The relationships between people were just so strong. It started from the top, from Henri."

The Genzyme J.P. Morgan event, which endures to this day, became one more testimonial to what was one of Termeer's greatest legacies, his development of an army of biotech leaders who would survive him. They had been developed to excel, innovate, give back, and introduce therapies for unmet medical needs, especially those directed at rare disease patients.

Over his long reign at Genzyme, Termeer circumstantially had become the ultimate biotech human capitalist. As one of his executives had said, "Henri wasn't born great." And the birthing of a next generation of rare disease industry leaders had not been part of his original Genzyme mandate.

But as his leadership skills, his stature, and his formidable executive presence developed, this significant contribution of a massive talent base to the biotech industry came into full view. He not only had had a nose for talent but also the awareness of its importance in the company's success. While Termeer was only its 17th U.S. employee, Genzyme's employee ranks peaked at nearly 14,400 in 2009. He had attracted top talent, nurtured it, and delivered a multitude of leaders who were highly sought for leadership roles by investors and life sciences companies. The result was his diaspora.

Peter Drake, the famed biotech investment analyst who often opined on such matters, met Termeer in 1984, and visited with him every six months for 15 consecutive years. His firm, Kidder Peabody, served as the lead underwriter of Genzyme's 1986 IPO. He had this to say on Termeer's leadership team, "I watched him make some of the best hires I think any CEO has ever made. They were impeccable. I dare say you won't find another organization that had a deeper bench of talent, and where more executives spend time, and go on to run other great companies. Maybe Genentech. He surrounded himself with unbelievably smart people, all of whom were incredibly loyal to him. It was a total family."

They were not only talented but considered among the elite. And they were for the most part, as Drake had recalled, beyond loyal. Voluntary turnover at Genzyme hovered between one and two percentage points during Termeer's 28 years. It was widely recognized as a great place to work. And the company's "Good Housekeeping" seal of approval inured to the reputational benefit of its executives. It was akin to certification.

These men and women had been trained under Termeer's watchful eye, developed, and in many cases, mentored by Termeer himself. When they left Genzyme, whether prior to or after the Sanofi acquisition, they were prepared to lead. He knew them all by name. He knew most of their families. And to each of them, he was forever just, "Henri."

People asked how would this enormous talent base be selected? And how would Termeer develop them? What was it about Genzyme that created this unusual concentration of exceptional talent?

Some would ascribe Genzyme's supportive culture as the foundation for not only the company's success, but also its executive talent factory. Although a subtly competitive place to work, its virtues were built around service to patients, trust, and the creation of an environment in which each employee could achieve his or her full potential.

Virtues, after all, are the states of the soul, and Termeer understood this in spades. His set of virtues, the highest of which was moral virtue, anchored Genzyme. The virtue of serving the well-being of humanity permeated every aspect of the company's mission. Into this supportive, some would assert "cultish" (others, "paternalistic") environment, Termeer would identify, attract, and mold many of the best and the brightest.

Many of Genzyme's top executives and the ranks below them came from careers in academic medicine, law, consulting, and other service industries that were typically reservoirs of talent compatible with Termeer's wish list. These places were not the usual sources of biotech executive talent. Alison Lawton had this to say, "We had quite a few very strong characters at Genzyme, people who did not necessarily comply with the usual way of doing things. But Henri understood that they valued self-initiative, accountability, and intellect."

They were also from organizations that tended to operate with flat organizational charts and often were characterized by ambiguous authority, less structure, and a healthy degree of institutional chaos. Energy and ideas were released in these environments; meritocracy reigned. By comparison with other major biotechs, fewer of Genzyme's top executives had had long stints with major corporations, excluding of course Baxter, which Genzyme mirrored culturally in so many respects. Termeer loved the lack of bureaucracy and sought to hire mavericks who would resist its encroachment.

He also liked the fact that professional firms were sources of free-thinkers. He liked intellectual nonconformity, especially when it was directed toward solving problems and changing the world for patients. Buying into the Genzyme mission when they joined the company, they would help him shape Genzyme's culture and its future.

Another cultural thread that would bind Genzyme's leadership was pace. Termeer's awareness of time played a substantial role in creating Genzyme's high-performance culture. Termeer collected clocks. Many had been transaction closing gifts from his bankers. Remembering Bo Piela's story of Termeer's 12th birthday gift from his father, Henri would live much of his professional life in a hurry. Why the rush? The answer always reverted to his unrelenting, urgent mission of saving patients. Every minute was precious, a patient's well-being hanging in the balance.

Although no official figure exists, at the time of his death in 2017, Termeer could lay claim to the fact that he had trained more than 100 CEOs who had cycled through the executive ranks at Genzyme prior to their ascension to the role. A high percentage of these sought to extend Genzyme's mission by pursuing new technologies that enabled the treatment of new rare genetic diseases. And at the time of his death, Termeer was mentoring 46 active CEOs. He was quietly proud of these achievements, and they speak directly to his greatness and leadership capacity, not to mention his legacy.

Jeff Albers, a young, mid-level development executive when he joined Genzyme in 2004, was an illustrative example of the company's diasporatic phenomenon. Having attended Georgetown University, Albers received a JD/MBA, after which he entered the private practice of law. Initially working closely with Peter Wirth on transactions, he had stayed with Genzyme through the completion of the Sanofi acquisition. He had been promoted into an oncology leadership role, and as a result, at a young age, gotten substantial exposure through these assignments to Henri Termeer.

Albers reflected on Termeer's impact on his outlook and advice, "Henri would start with the people component, including employees and patients. He had a very caring nature. He would also get to the few key questions very quickly. He understood that we were in an environment that was evolving, so we had to keep pace and evolve with that. And his view was that we could always be better, continue to grow professionally. He was shaped by a generous spirit—these stood out beyond all else. I often find myself now as a CEO, thinking 'How would Henri have dealt with that?'"

Albers sits today atop a leader in personalized medicine, Blueprint Medicines, which he joined as CEO shortly after its 2011 founding in Cambridge, Massachusetts. The large rectangular platinum sign in the company's lobby announces in vertical succession the company's five core values. "PATIENTS FIRST, THOUGHTFULNESS, OPTIMISM, URGENCY, TRUST" it reads. For all one would know, they were visiting Genzyme II.

As he began building out Blueprint, Albers relied on Termeer for commercial, cultural, and strategic guidance when important decisions were on the line. He would often seek Termeer's counsel privately to complement his board's. He was one of Henri's 46 mentees, and as much as Albers treasured this relationship, so too did Henri.

Another mentee was John Butler, who had spent 13 years at Genzyme, starting there in 1997. Today, as President and CEO of Akebia Therapeutics, a renal disease company in Cambridge, he had run Genzyme's renal franchise from 2002 to 2009, followed by a two-year period in which he ran its rare disease business. Pictures of Termeer and patients decorate his office.

As Akebia's leader, Butler reflected on Termeer and the lessons he learned from the master, "What I learned from Henri is you can turn around, but there isn't anybody else, right? There's nobody else. It's on you. Whether it's a manufacturing problem, a clinical trial issue, or a commercial issue, it is ultimately your responsibility. I learned that from him.

"Henri did have that ability to dive in, in a way that is crazy... . You know, asking the questions he would ask ... you couldn't snow him. He knew. He had the computer screen open. He knew exactly how many vials we had, and you damn well better know that too. He would question decisions and we would discuss them, but he would always be supportive. Never once did he override a decision on me. He put you in charge of the business with a board of one—Henri."

Butler confirmed a Termeer rule that others often expressed. On taking up a general manager's role early in his Genzyme career, Termeer had admonished him, "John, you can disagree with me anytime, but never do it in public."

Many of Termeer's mentees likened their relationship at Genzyme with Termeer as that of two business partners, with Termeer the senior partner and the mentee as junior partner. The dynamic was unstated but obvious. It was a construct that fostered trust, intense loyalty, and alignment of mission, which in Butler's words was twofold, "a maniacal desire to serve patients and a maniacal desire to hit EPS on a quarterly basis."

It also enabled Termeer to stay close to the business, while enabling his progeny to sit at the knee of the master engaged in his craft. It was indeed a form of apprenticeship. And as Greg Phelps offered, "It was a really motivational way of managing people. That was as close as there ever was to any sort of official mentorship."

David Meeker, who for 23 years had been developed by Termeer and became his immediate successor as Genzyme's CEO, had a different way of describing Termeer's mentorship style, "Henri was a great model in the mentor sense. We all learned from watching, not from coaching. I got very little coaching from Henri. Yet, he knew he was developing people without making a big deal of sitting down and coaching people. For me, it was modeling. He personally made sure I couldn't get enough opportunities to watch." Meeker and others described Termeer's style of mentoring as "somewhat informal." Some would visit Termeer with formal agendas, but most of the visits were unstructured chats.

Once Henri's mentee-leaders had left the Genzyme nest, they knew the discharge of their next assignment was their responsibility. For Termeer was famous for ending discussions with his colleagues in three words, "It's your responsibility." The dictum was repeated in conversation after conversation. In his worldview, personal responsibility and accountability were nonnegotiable.

Another facet of Termeer's diaspora was his commitment to the development and promotion of women. Beyond Gail Maderis, who became an exemplar of the talented women whom Termeer would mentor and go on to lead organizations, there were a host of others. Other Genyzme alumnae to become public company Chairwomen or CEOs would include Alison Lawton, Paula Soteropoulos, Julie Anne Smith, Ann Merrifield, Alison Taunton-Rigby, Paula Ragan, and Alicia Secor. All are Genzyme alumnae; Termeer played a role in the development of each of these talented leaders. In several instances, he played a major role.

Alison Lawton is as good an example of how Termeer would develop and mentor talent as exists. She had joined Genzyme in 1991 as a lowly International Regulatory Affairs Specialist, having emigrated from London to Boston as a trailing spouse. Today she stands astride Kaleido Biosciences, a biotech upstart unlocking the power of the human microbiome, as its CEO.

The story of Termeer's 26-year mentorship of Lawton, 20 of which were during their years together at Genzyme, is instructive on many levels. But the important lessons are those that relate to his willingness to take risks on people, his ability to infect those around him with his optimism and can-do mentality, and his ability to get the absolute most out of his people.

Lawton, who was ultimately selected in 2010 to lead Genzyme's $1 billion biosurgery business, commented, "Henri loved to develop people and give them opportunity. If somebody was doing an amazing job, he was willing to take a risk on them, especially if he was convinced that they were the right kind of talent. He was a huge proponent of giving people stretch development opportunities.

"With Henri, there was always this sense of urgency around the patient. Henri was a real risk-taker and he would push. If he thought things weren't moving fast enough, he would challenge that.

"He also protected his champions, especially the country managers, giving them autonomy. His belief was that they know their country, they know patients in that country, they know what's right for that regional part of the world. Allow them to make decisions rather than being driven from corporate.

"But he also believed in having a matrix, because it created tension in the system, and by having people driving and pushing, checks and balances were created. You'd end up in the right place."

Another woman Termeer would mentor was Paula Soteropoulos who today leads Akcea Therapeutics, a publicly held rare disease company focusing on RNA-targeted medicines. Soteropoulos joined Genzyme in 1992 as an engineer. Her first role was managing the build-out of a manufacturing site in Framingham, a suburb of Boston.

Soteropoulos' career was propelled by Henri Termeer who recognized early on her raw talent and capacity for growth. Her story was like that of so many others who joined Genzyme at the beginning of their professional development. She had gotten direct, regular exposure to Termeer in her second year at the company, managing Genzyme's global capital expenditure budget—more than $100 million.

Like Ed Kaye and others, she would stand up to Termeer in managing her responsibilities if she felt justified in doing so. She recalled, "Henri didn't like it, but he appreciated the fact that I challenged him and pushed back. I was new to Genzyme and many layers down in the organization. He was frugal with money. He taught me early about accountability and how you find new innovative ways to get things done on a tight budget."

Over the years, as her leadership capacity became more obvious and her confidence grew, his schooling intensified. Her assignments rotated frequently, each one building on the other. By the early 2000s, they would have monthly one-on-one career discussions at his Genzyme Center office, always free-flowing, hardly an agenda. Soteropoulos remembered, "He'd often open the conversation with 'Okay, what's on your mind?'" She further recalled, "He was the consummate listener," a skill she has never forgotten.

"I personally never had aspirations to be a CEO. I never thought I could or would be. But Henri believed in me more than I believed in myself. He recognized that women often hold themselves back. One day, he asked me, 'What do you wanna do, where do you wanna go? I think you should be a general manager.' Henri opened that door for me.

"Henri took risks on people. He did that with so many of us. One of the single most important things in my life is he took a risk on me. He trusted people, he took them out of their comfort zone, and sometimes it backfired, but he would hold you accountable and performers could shine."

He also gave Soteropoulos lessons and courage in pricing the expensive medicines that treat rare diseases, "He never apologized. He deeply believed that it was his responsibility to drive innovation, and as part of that, you had to build a sustainable business. He believed it was our responsibility as a company, but it was also our society's responsibility. He would say 'If we're going to fund the next generation of innovation, we have to command a certain price and a profit. And we have shareholders who expect a certain return. It is hard. I know that it is hard, but that's why we're doing this.'"

Scores of other Termeer mentees would go on to lead companies using new life sciences technology platforms that address rare diseases too complex to have been treated during his tenure. Many others would sit on the corporate boards of rare disease enterprises. The common thread would be hope—providing hope to rare disease patients, many of whom were afflicted with diseases too new or too rare to have been addressed by Genzyme.

The beautiful irony of Termeer's diaspora is that it would never have happened were it not for Sanofi's acquisition of Genzyme. Termeer had resisted it. But as David Meeker, reflected on it years later, "The acquisition paradoxically saved Henri. It gave a certain eloquence to the ending."

What followed ensured Termeer's place as one of the greatest biotech CEOs ever, for the deal had unlocked powerful forces, revealing much of his life's impact on the future of rare disease therapy development. Gradually, waves of executives would depart the company. Termeer's protégés would seed the next generation of innovation. He had inspired them and provided each with the skills to deliver on the magic they would create and a mission that many would carry forward.

Termeer's generosity in supporting the careers of the men and women of Genzyme's diaspora was perhaps unparalleled in biotech history. The breadth and commitment of his legion of acolytes to extending the rare disease movement he started—all in the service of fulfilling the responsibility he personally felt to the well-being of individual patients—reverberates today as a core and profoundly important part of his legacy.

CHAPTER THIRTEEN

A Citizen in Full

It was 8:00 p.m. on Friday, October 29, 1999. A glittering audience of prominent Bostonians, biotech leaders, and friends of Henri Termeer were gathered in the ballroom of the Royal Sonesta Hotel in Cambridge for the year's presentation of the Golden Door Award. The International Institute of New England, which had been providing services to immigrants and refugees since 1924, had been giving the award since 1970 to recognize "the positive influence that immigrants have had on this country by honoring a distinguished American of foreign birth."

The award's name came from a poem inscribed on the base of the Statue of Liberty, "I lift my lamp beside the golden door." Previous recipients had included I.M. Pei, Arthur Fiedler, and Yo-Yo Ma.

Tonight, the award would be bestowed upon Henri Termeer. But first, to fulfill the award's other qualification—that of American citizenship—his naturalization ceremony would be made a part of the program and precede the award's bestowal.

Standing on the ballroom stage, dressed in black tie, the five-foot-ten, 170-pound Dutch candidate stood at attention, with his left hand raised and right hand across his heart, awaiting the officiant's instructions.

Former Massachusetts Governor Michael Dukakis, the evening's Master of Ceremonies, instructed the INS agent to begin. He would lead in recital, "Repeat after me ... I hereby declare an oath ... ." Henri Termeer would complete a journey he had begun 28 years earlier and consummate 140 words later. He would leave the dais as a U.S. citizen.

Nineteen days later, Senator Ted Kennedy entered his own tribute to Henri Termeer in the Congressional Record, congratulating him not only for his Golden Door Award but his new citizenship.

Now 53 years of age, adopting U.S. citizenship had been a decision Termeer had weighed for years. Should he formally become one? He was and always would be proud to be Dutch. But the pragmatist within him had finally prevailed. He had developed a deep appreciation for the United States, and by now, he concluded the time had come.

The year before, Termeer had brought another personal journey to a close. His first marriage had dissolved in divorce nine years earlier, and his first wife had long ago returned to her native United Kingdom. He dated several women in the intervening period, but Belinda Herrera was the one who would become his true soul mate. She loved and supported him as no other.

Belinda was a New Mexico native who had resettled years earlier in Boston. She had worked first at Integrated Genetics (IG) in 1985 as the Financial Administrative Assistant. At times, she was acting as temporary assistant to IG's CEO Bob Carpenter, another biotech pioneer and close friend of Termeer from Baxter. When Genzyme acquired IG in 1989, Carpenter stepped down and Belinda joined Termeer's office as Executive Assistant.

Warm, smart, and beloved by her colleagues, she and Termeer quietly dated and would over time become romantically involved. In recognition of the changed circumstances, she chose to leave her position at Genzyme in 1997 without fanfare and moved to California, staying with her sister's family. But by then, she and Termeer had gotten to know each other heart and soul, in sickness and in health, until death would do them part. Henri would ask for her hand in marriage.

In a small ceremony held in the living room of their oceanside house in Biddeford Pool, Maine, their August 1998 wedding was attended by four couples and a few family members including Termeer's son, Nicholas, who would travel to Maine from the United Kingdom.

After the wedding, Termeer, ever the Dutchman, took Belinda on a camping trip in the nearby White Mountains of New Hampshire. It was to be their "honeymoon."

It was some honeymoon. Nicholas joined them, and they hung out at a campground for a few days in a small blue nylon tent, swatting mosquitoes and grilling hot dogs. Nicholas slept between them. With Henri now earning princely sums as the CEO of a major, world-class, public biotech company, Belinda might have conjured a two-week trip to Tahiti or some such, more fitting romantic destination. Instead, it was to be a camping trip to the White Mountains with his 12-year-old son in tow.

But she was not one to complain. It was no surprise. She knew exactly the man with whom she had fallen in love and married.

In October of 1998, Henri and Belinda would purchase a harbor-front house in Marblehead just around the bend from the Corinthian Yacht Club. They had been residing in a small pied-á-terre on Commercial Wharf, in downtown Boston, overlooking the harbor. Within two years, during the summer of 2000, Belinda would give birth to their only child, a beautiful daughter, Adriana.

By now, Termeer was deeply ingrained in Boston's social whirl. Although he still spoke with a Dutch accent, even a heavy one at times when it could be invoked for greatest effect, he preferred to speak English. More than just a *lingua franca,* it had long ago replaced Dutch as his principal means of communicating. On his frequent visits to the Netherlands, if other Dutchmen spoke it, he was known to ask if they would instead mind conversing in English.

Termeer was not a socially ambitious person, but he was one who enjoyed spending time with other leaders who shared his intensity and large vision for the future. And his interests and vision were not confined to biotechnology. Over his remaining years, he would demonstrate a refined and growing interest in the arts, education, economics, and the delivery of healthcare to the community at large.

"Time is the stuff of which life is made," another wise American innovator of another generation, Ben Franklin, once mused; and so it was for Henri Termeer. He too was acutely aware of his time and how he would spend it. His interests outside the office often connected to his official duties, and they were a priority in how he lived his life.

They also became platforms that contributed to his reputation's growth and importance. He would become known as an exceptional, grounded, wise leader. The potentates of Boston's venerable institutions and biotech boardrooms would seek his advice, decorously welcoming him into places known to be populated with a swelling ego or two. He had the grace, wisdom, intellect, and sense of humor to fit in.

Some would say that Termeer, too, possessed an initial, reserved aloofness of his own, but on meeting and spending time with him, it melted away as his considerable sensitivity and humility were revealed. He struck almost everyone as naturally likeable. The twinkle in his eye, his willingness to listen to others' points of view, and his natural charisma were persuasive, endearing assets. He was one who could make connections easily. Although Dutch, he may as well have been from Switzerland; his appeal would become universal.

Among the first projects in which he invested his time was the establishment of Boston as a global life sciences center. By the late 1980s, several years after he had joined Genzyme, the greater Boston area was being challenged, most notably by San Diego and the San Francisco Bay Area, as the global epicenter of biotech.

Kendall Square in Cambridge, which would become the most densely populated concentration of biotech enterprises and talent in the world, was then but an embryo of its later self. San Francisco had Genentech; San Diego had Hybritech. Genentech was in a class by itself as the industry's leading company and talent factory. Hybritech was the first of the major biotechs to be acquired (in 1986 by Eli Lilly for $400 million), spawning talent and a host of new energetic, well-funded start-ups.

Termeer was not one to accept the status quo and would fill a vacuum, emerging as a leading spokesman for the promise and attraction of Boston as the world's most prominent life sciences ecosystem. In 1985, this crusade led him to cofound MassBio, which would become an advocacy juggernaut that today represents the interests of 1,200 member life sciences companies.

Later, one of the large international strategy consulting firms would convene a blue ribbon committee, populated by not only Termeer but other biotech executives and stakeholders, to create a blueprint for Boston's life sciences future. Termeer had just taken Genzyme public and his interests in the committee's recommendations and action plan were considerable. The path he and others laid would transform Boston.

In the early 1990s, Termeer would join the Board of the Museum of Science, one of Boston's preeminent community institutions, and would begin a string of affiliations with Harvard's medical institutions. The first such affiliation was with a Harvard Medical School (HMS) working group, otherwise known as "Joan's Fantasy."

At the time, Dr. Joan Reede had recently joined HMS to lead the school's diversity programs. She had been a young pediatrician employed as the medical director of a Boston community health center, supporting children in public schools and juvenile prisons. She had this idea. She fantasized the creation of a program to identify, support, and mentor minority students, trainees, and professionals so that they could pursue otherwise unknown biomedical and health sciences careers.

Reede described her challenge and how her first encounter with Termeer would change her life, "I was naive. I was a newbie within the HMS environment. This was a whole new world for me. So, I wrote all these companies. Because the companies were always saying, 'We care about diversity.' I only got one response, from Genentech. I was in my early thirties and frustrated.

"So I was speaking with someone who described Henri as the most powerful, the most influential, the most successful person in the biotech/pharma world within New England. I picked up the phone and called directory information, 411.

"I called Genzyme and asked for Henri Termeer. I was simply astonished. He came to the phone. Henri took the call, not necessarily knowing what was going to be on the other side, but willing to give me a chance. He listened for 15 minutes. He came on board. When I think of Henri Termeer, I think of someone who would take a call."

In collaboration with the Massachusetts Medical Society and the New England Board of Higher Education, Dr. Reede soon thereafter founded the HMS-sponsored Biomedical Science Careers Program (BSCP). Over the next two-and-a-half decades, Termeer became one of Reede's staunchest supporters, not only providing BSCP with financial support, but also encouraging, advising, and mentoring this young leader. He guided her in forming a 501(c)3 and joined her board. His assistance was instrumental in enabling her to build one of America's most successful minority life sciences diversity programs. She was appointed Dean for Diversity and Community Partnership at Harvard Medical School in 2002.

Although it would be the first, Termeer's connection and deep involvement with Harvard and its medical institutions would hardly be limited to Reede's successful initiative. In 2004, Termeer would join the Board of Trustees of the Massachusetts General Hospital (MGH), Harvard's flagship teaching hospital consistently ranked as one of the top hospitals in the world. And in 2008, he would join the Board of Directors of Partners HealthCare, the overarching, HMS-affiliated healthcare system that would provide medical services annually to thousands of patients through its five prestigious Harvard teaching hospitals, including MGH.

Dr. Peter Slavin, CEO of MGH since 2003, remembered meeting Termeer at a symposium in the early 2000s at Harvard Business School. Termeer had been invited by Harvard President Larry Summers and the event's cohost, Michael Porter, a distinguished HBS faculty member, to comment upon what could be done to maximize the impact and success of the life sciences assets in the greater Boston area.

He recalled, "There was a series of speakers and open discussions over the course of the day. There were probably about two dozen people who spoke or participated in panel discussions, and one of them was Henri Termeer, whom I had never met... . I found his comments were light-years ahead of everyone else's, and I was blown away by his intelligence, his insights, his charm, and his articulateness. I left the conference wanting to know more about him. I wanted to get him involved here at Massachusetts General Hospital.

"Some of the best trustees for nonprofit organizations—any organizations—are the ones who give you insights into your business that you otherwise would not have had. Henri was a master of that."

Dr. George Daley, a renowned stem cell biologist and Dean of Harvard Medical School, had a similar reaction. Daley would meet Henri Termeer many years later on his being named, in 2017, the new dean of the august institution he now leads. Harvard President Drew Gilpin Faust would host a dinner at Elmwood, the historic home of Harvard's presidents, to introduce the incoming Daley to various leaders of the life sciences community. Termeer was an invited guest.

Daley would cogitate that spring on a way to get him more deeply involved in the HMS community. He had served on the HMS Board of Fellows since December 2000. He would meet with Termeer a few weeks before his passing. Sadly, it would become one of the many unfinished conversations that accompanied the unexpectedly sudden death of this fallen leader.

Daley would later remark, "I think there are very few leaders who enjoy the universal respect of someone like Henri. The universality of his respect was remarkable. He reminds me of my PhD thesis advisor at The Whitehead, David Baltimore, who won the Nobel Prize at a young age, 37. Like David, Henri enjoyed great early success and was able to adopt a personality of generosity that is unusual. Henri used his power and his reputation in a very beneficent way."

Another Harvard icon, Dr. Samuel Thier, former CEO of MGH and Partners HealthCare, saw other ways Termeer would contribute as one of Boston's preeminent citizens—he was a leader who would pull people together.

He would remember Henri's contributions in this way, "Henri had a wider view of the world than most. He didn't see things in a transactional way. He could see the big picture. When he died, it was as though a whole piece of our health community had died. He was such an important, central figure. He was gifted in bringing people together. His loss left an enormous hole. We had always counted on him as a critical part of our local culture and competency."

Another member of the Harvard family, Dr. Daniel Haber, would also reap the benefit of Termeer's generosity. Termeer and his wife would make a founding gift in 2011 to the Massachusetts General Hospital, providing the impetus behind the creation of the Henri and Belinda Termeer Center for Targeted Therapies, a clinical research center of excellence focused on personalized medicine. As a core part of MGH's Cancer Center, led by Haber, the Center would serve as a reminder of Termeer's beneficence.

Haber remembered Termeer for his vision and the center's fit with Termeer's strategic approach to medicine, "It was a perfect fit for Henri because this was going from rare diseases to rare cancers, finding niches where you can make a really important difference in a subset of the population... . Initially, the Center was just going to be a footprint. And then, something magical happened, which is that it became the center for pretty much all of our early phase drug tests in the cancer arena at MGH." It would also become a center where promising drug therapies would be tested.

While Termeer was building his commitment to Harvard's various clinical and research institutions, he concurrently devoted himself to the advancement of science and medical research innovation through his service as a Member of the MIT Corporation, the governing body of the Massachusetts Institute of Technology. Becoming a Member in 2006 at the invitation of Chairman Dana Mead, Termeer would serve 11 years until his passing. In 2011, Termeer was invited by then–Corporation Chairman, John Reed, to join the Executive Committee of the MIT Corporation.

Rafael Reif, PhD, was serving as Provost of MIT in 2006. An electrical engineer noted for his contributions to microelectronics, he would be selected in 2012 to serve as the Institute's President. Termeer would serve on the search committee that recommended his selection.

Reif had witnessed firsthand Termeer's influence on the MIT board, which he would characterize in two particular ways—his calming demeanor and his curiosity, "It is remarkable to see an individual who is so calm, so thoughtful, and so wise in the sense that there are so many people on boards who like to show how smart they are in very many ways. Henri was the calmer one. He just listened, never got too upset or too happy. He was always very steady. And when he would feel that the conversation may go a little bit off mark or could go in the wrong direction, he would just make an observation and very gently put everybody back on the right track.

"Henri was a very, very honest, decent, fair person. I never saw any hidden agenda. He just was there to help the institution be the best it could be. He was always thinking of what was the right thing to do. Henri was somebody who when he had something to say, everybody would stop and listen.

"He was extremely curious about all sorts of possibilities. Clearly passionate about biomedicine, anything dealing with biology, health sciences, or technology, it was like a magnet for him. It wasn't like he was curious about astrophysics and galaxies, but if he saw a connection, he would be curious. In that sense the universe was his curiosity."

Dr. Susan Hockfield, Reif's immediate predecessor and a new arrival from Yale in the fall of 2004, would remember Termeer in this way, "I met Henri early in my presidency of MIT. I did not appreciate the power of the region. Henri was fantastic at describing the region and how we could use it to amplify our work in the world beyond. He was a real advocate, and it was so contagious."

Hockfield, a neuroscientist and the institution's first female as well as life sciences President, would describe Termeer's influence on the development of Kendall Square, "When I arrived, Kendall Square was not what it is today. Novartis was here, but Pfizer had not yet moved in. There were a lot of building sites and parking lots, but it wasn't built out the way it is today. He was a founder, one of the main founders of the twenty-first century Kendall Square.

"I don't mean it in a political sense, but Henri was also enormously supportive of the advancement of women. I kept running into women who were in various significant positions at Genzyme, and we all know that biotech is not the friendliest of places for women. He was remarkable in advancing the careers of many women."

The themes of thoughtfulness and generosity touched Termeer's service to the array of institutions in which he became involved. Prominent among Henri's passions was the Boston Ballet. Inspired and founded in the 1960s by E. Virginia Williams, the ballet would become one of Boston's most acclaimed cultural institutions. Its current Artistic Director, Mikko Nissinen, a Finn, had immigrated to the United States in 1987 and would lead the organization, his self-acclaimed, "Ballet of the Future."

Nissinen would become one of the ballet world's most forward-thinking artistic directors, taking his troupe around the globe and leaving its mark from Paris to Seoul. He would also build on Williams' legacy of training young dancers, substantially expanding its ballet school, an interest he and Termeer shared.

One of Nissinen's students would be Adriana Termeer, and the Termeers became devoted supporters of the Boston Ballet in the process. First Belinda Termeer, then her husband, would serve on the institution's Board. They also assisted in establishing a striking, new Boston Ballet outpost in North Boston, near their home in Marblehead.

Nissinen recalls Termeer as a dreamer who got things done, "He was the dreamer who was not afraid to dream big things. But he wanted to get things done, and then move on, and then do more. It was like he cut through the BS and said, 'Okay, this is how I see it,' and he was bang on."

He also saw in Termeer an exceptional capacity, rooted in his listening skills and Dutch humanism, to provide others the space they would need to reach their full potential. Nissinen would remember, "I have to say with all the greatest individuals that I've met in my life, he possessed the quality of being a fantastic listener. Yes, he could speak. His gift of speech was golden, but his gift of listening was priceless.

"Henri was so kind and, you know, accepting. He was the total opposite of those who try to figure you out and put you in a box. He instead created more space around you, so you could be yourself, expanding rather than limiting who you are.... He allowed you to be larger. He would not diminish you or define you. He expanded the box to the point where there was no box.

"If you listen to some of the Buddhist chants, they were in his tone of speaking and manner of being. They were embedded in there, like a deep 'Ommmm.' He just let it be. It was him."

As the ballet's artistic director, Nissinen's insights would prove to be unique, but no less predictive as to how Termeer would handle the stress of the financial crisis that would befall the world economy in 2008–2010.

Invited to join as its Deputy Chairman, Termeer would commence four years of service as a Member of the Board of Directors of the Federal Reserve Bank of Boston in 2007, a year before the global economy would nearly collapse. In accordance with the Federal Reserve Act, he would be one of three Class C Directors who would be asked to serve to represent "the public interest" on this, one of the twelve Federal Reserve banks that are collectively responsible for promoting the nation's economic growth and financial stability, as well as protecting its banking system.

Eric Rosengren, President of the Boston Fed, would remember, "Henri was brought on primarily because the pharmaceutical industry is a key industry, and health and life sciences were becoming increasingly important to the New England economy... . As we got into 2008, it became clear that problems with the economy were getting much worse. Bear Stearns failed. Interest rate spreads were widening. And as we got into the fall, Lehman Brothers failed.

"The Boston Fed and the New York Fed were the only two reserve banks that had liquidity facilities during the crisis. The Boston Fed's facility was focused on stemming the run on the money market mutual fund industry. We designed something over a weekend, the AMLF [Asset-Backed Commercial Paper Money Market Mutual Fund Liquidity Facility], a fund to provide liquidity, and it was highly risky, but successful in stemming the crisis.

"Henri understood the severity and the complexity of what was going on. He was a very calming influence on our Board. It was a time where it could be very easy to get quite disturbed."

By late 2009, Genzyme was in a crisis of its own, and Rosengren reflected on Termeer as he toggled between the nation's economic crisis and the one in bloom at Allston Landing. Termeer had just been named

Chairman of the Boston Fed Board that November, with effect on January 1, 2010, "I think it was a very tough period for him. He was serving as our board chair at the same time that he was dealing with very complicated issues in his own company. He didn't really show a lot of stress that I'm sure he was feeling at the time. If you were an outside observer just sitting in our boardroom, you would not have known that anything had changed in his life. He was able to maintain a very tranquil demeanor, despite the fact that I'm sure his life was being completely upended."

Some years later, as other national and civic leaders would reflect on Henri Termeer's life and the many not-for-profit institutions and philanthropic causes he served, none could hardly sum up his service better than Massachusetts Governor Deval Patrick.

Patrick had met Henri Termeer in 2005, as an early, long-shot candidate for the Massachusetts governorship. He was one of three Democrats vying for his party's nomination. They had met at a biotech conference.

Governor Patrick remembered Termeer for his kindness and openness in greeting him. It was an encounter that would lead to a 2005 invitation to visit Genzyme's newly constructed, Leadership in Energy and Environmental Design (LEED)-certified headquarters building. It was also one that would provide the foundation for a friendship that would endure for the balance of Termeer's life.

Deval Patrick would go on to be elected the 71st Governor of Massachusetts in 2006, succeeding Mitt Romney and taking office in early 2007, serving two four-year terms until early 2015.

When asked to summarize Henri Termeer's life and his impact on society, Governor Patrick would look down, collect his thoughts in a moment of silence, and return to eye contact, describing Termeer in these terms, "I think of Henri Termeer as a 'citizen in full.' What do I mean by that? It's back to this notion of civic presence. He was, yes, a successful and consummate businessman. And, yes, he was a loving husband and family man. But this notion of being a member of a whole community, and contributing as a member of the whole community, seeing your neighbors' dreams as well as your own. This is how I remember Henri.

"He was of this place, and it was of him. And the notion that he had a contribution to make, and that he was listening to the contributions others could make to the community, that it wasn't just him with the answers. He was humble that way.

"As gifted as he was, he was really interested in what others had to offer. And that's an extraordinary thing for a powerful man to bring to bear. He brought out the best in others."

CHAPTER FOURTEEN

The Oracle of Marblehead

The evening's dinner had concluded at the Algonquin Club. Located in the heart of Boston's historic Back Bay neighborhood, with its grand staircase, soft lighting, and elegant furnishings, its walls whispered tales of years gone by. Termeer had long held board dinners and other Genzyme events at this, one of his favorite venues for gatherings of this sort.

Tonight's program, however, had been different. "Bittersweet" as some would later describe it. The event had pretended a celebratory air. But everyone knew, inevitably, things would never be the same.

Seven tables accommodated the 52 who gathered that night for a round of "goodbyes." Some had traveled a considerable distance to be there. It was a night to reflect, remember their accomplishments, and savor their years at Genzyme. It was also a night to mourn their loss.

The Sanofi acquisition would close in the next week. "It was a little sad," recalled Belinda Termeer, "because this was something that Henri did not want to happen. He really wanted to end his career with Genzyme on his terms, and the unfortunate thing, the virus, and everything that happened afterwards ... I know he felt proud of what he had accomplished, and he took the responsibility for what happened."

With a few exceptions, those in the room had been or were Genzyme board members and senior officers. Several of the alumni, including Henry Blair, had been there in the 1980s, adding warmth and depth to the storytelling.

Some in the room who would continue their careers with the company's acquirer—David Meeker being one—would also attend. Sanofi and Chris Viehbacher had made considerable effort to make them feel welcome.

Elliott Hillback was not one to miss its significance. He had set up a tripod with his video camera, capturing the speeches and toasts as every attendee spoke and the honors were passed around the room. "It was a rare night," he recalled. "Pretty emotional."

At the end, Termeer delivered a few, passionate remarks. His peroration thanked all for their loyalty, their dedication, and their effort. He was visibly somber in delivering his message. They had done all they could to serve orphaned patients, build a unique company, and maintain control of their destiny. Now, it was time to move forward.

At this juncture in his life, Henri had his health and the financial independence to do just about anything he wanted. He would remain on Sanofi's books through midyear. Most then expected him to retire.

What no one could have foreseen, however, was that Termeer had but six precious years left to live. Having just turned 65, he was as vigorous and curious as ever. Although in need of a short break to recharge and recover from the turmoil of the past two years, he was at or near the peak of his capabilities and aptitude.

He took leave of Back Bay that evening, crossing the Tobin Bridge en route to Marblehead in his black Mercedes with Belinda, and he had, in his own way, begun turning his attention to the future.

After all, Henri's life had always been about the future. He was never one to dwell in the past.

Phil Sharp, PhD, the noted 1993 winner of the Nobel Prize in Medicine or Physiology, Professor at MIT, and cofounder of Biogen, knew Termeer well. "He was never interested in looking backwards," Dr. Sharp remembered. "He was only interested in looking forward. He was always in a hurry. He saw so many things he could do to improve lives, helping others."

Belinda, too, understood this aspect of Termeer's makeup. She had seen it ever since they had met in the late 1980s.

But, on this particular night, there was another side of her husband's state of mind that had caught her attention. She had never seen this side of him before. His self-confidence had been shaken. "This was a very painful time for Henri. It was an emotional period. He was strong at work, but when he came home, I think the biggest fear for him was that everyone would see this as a failure, not a good thing for the company... . It was a tough time.

"I think he was a little nervous about what was next. I think I was more worried for him than he was, I don't know. Henri wasn't someone who could easily relax. He needed to keep himself busy. He couldn't sit home all day.

"I tried to be a calming presence, reassuring him that he did the right thing, and he was going to be fine, and as I told him, 'People will want you, I know you're going to get calls.' But I think he was really nervous that people would forget about him and not want his involvement because he still wasn't finished realizing his big dream.

"His resilience, even through stressful times, helped him thrive despite the challenges. I never met anyone like Henri who could have the most difficult challenges to deal with and be able to switch and become fully present in the moment on others. He always seemed to take the positive view from adversity."

Then, weeks later, people started calling. This "somber period," as Belinda would later describe it, would be short-lived, lasting only several weeks.

Not only would she and his family lift him through it, but so too would others. Some of his closest Genzyme colleagues had reached out to him, especially his Marblehead neighbors, Ed Kaye, Sandy Smith, and Jan van Heek, who were among the first.

But by June, it was over.

"I don't think it took him long to snap out of his uncertainty and nervousness," Belinda recalled. "It seemed like overnight he went from 'not sure what I'm going to be doing' to 'busy as ever.' More people came to see him and companies wanted his advice. Henri had an incredible gift of listening and communicating. He was mentally prepared and anxious to move forward. He saw his time at Genzyme as a learning and growing experience."

All of a sudden, it was as if he had become a clairvoyant, all-knowing, all-seeing maharajah. Ed Kaye nailed it. He would later, after Henri's death, dub him, "the Oracle of Marblehead."

In his humble way, it is fairly certain Henri would have loved it, for we all treasure the help we receive in times of doubt. But for Termeer, it was a lot more.

What became an outpouring of requests for his advice and counsel would serve as a reaffirmation, a reassurance that he could continue to dedicate himself to the things he loved—his work, the well-being of patients, and helping others. People understood the toll that the last two years had exacted upon him, and they were prepared to stick with him, through thick and thin. But they also recognized the deep well of experience, contacts, and wisdom he had accrued over the course of the last 35 years.

Not only would these sustaining activities continue, but he could also pursue new interests and ventures he had put off for years.

Three years prior, Henri and Belinda had purchased a graceful, three-story, yellow clapboard house next door. It had been built in the 1890s, had wonderful old bones and was set on a small bluff with a beautiful view over the Marblehead harbor. A stone wall had been built in front to protect the property from the nor'easters which pummeled the Marblehead shoreline each winter.

Henri approached his former Genzyme assistant, Mary Ellen Mucci, to ask if she would help him set up an office. There was room for the 200 or so boxes of material he had saved and plenty of meeting space for visitors. Mary Ellen obliged, and she and Belinda set about the task of freshening the house. It became known as "the Yellow House," and it would lend itself, with near perfection, to the next phase of Henri's life.

Termeer's last day at Genzyme was officially June 30, 2011, but right through his final month and beyond, Termeer was career counselor and mentor to hundreds. They would see Henri at the Boston Biotech CEO conferences, J.P. Morgan, Laguna Niguel, and events at Harvard or MIT.

He would pull people over to ask them how things were going. He would be consulted for his views on business plans and strategic decisions. He would often be asked for his views on the future, a topic on which he took great pleasure in expounding.

As one Genzyme executive, Zoltan Csimma, would put it, "Henri was very good at reconnecting with people after they'd left the company. You could leave the company, but you couldn't leave the family. He did not make long-term enemies with a very few exceptions." Termeer had a way of remaining in contact with his flock. If you had been a Genzyme employee, you were deemed to be a part of his family. And this pertained to not only those who were at Genzyme upon its combination with Sanofi, but those who were there in the earlier years.

As the former leader of the acquired company, Termeer would split his loyalty—to Sanofi, the new owner of Genzyme, and to his former colleagues who were assessing their career options. But this was a common state of affairs for departing CEOs, and Termeer did his best to support

Chris Viehbacher and David Meeker, while concurrently serving as a sounding board and mentor for those who remained employed at Genzyme.

Following his exit, Termeer was careful to avoid interference in Genzyme's affairs as he believed that Viehbacher and Meeker were able stewards of the company and its direction. The two of them worked well together and shared a vision for the combined businesses.

In Meeker's words, "Chris and I had a shared vision to allow this tiny island of culture to be a point of change within Sanofi, to be a point of influence and to bring in that simple thing that Genzyme has—an ability to show that a corporation can care about patients. It was such a fundamental thing." That, of course, had been a core premise in combining the two organizations. Meeker got it; he always had. But perhaps more importantly, so did Viehbacher who assumed ultimate responsibility in combining the two firms. Meeker later confided, "Chris really worked to understand Genzyme. He and Henri had developed a relationship. It was short and unique. But it was key to the success in integrating the two firms."

Termeer would spend his final six years taking great satisfaction in helping others achieve success and realize their dreams. His responsibilities in building Genzyme and serving its patient communities had been transitioned to a new owner, but his focus on supporting the growth and development of people who shared his passions never abated.

The list of those who sought his advice during this period was a long one, and yet, he seemed to go to great lengths to support them all. His efforts were by no means limited to counseling former Genzyme colleagues, and they stretched to the West Coast, Europe, and Asia.

Many were entrepreneurs and innovators with an interest in rare diseases. Some were persons who knew of his interests and experience in advising other biomedical pioneers. A number of the CEOs he counseled were of European descent. Among them were highly accomplished UCB CEO, Roch Doliveux (a Frenchman living in Belgium leading one of Europe's larger, well-regarded multinational pharmaceutical enterprises), Daniel de Boer (Founder-CEO of ProQr Therapeutics in the Netherlands), Elisabet de los Pinos, PhD (native to Spain, Founder-CEO of Aura Biosciences in Cambridge), and Kees Been (Dutch, Founder-CEO of Lysosomal Therapeutics in Cambridge).

Others representative of those with whom he met in plotting their futures or advising on leadership matters included Christoph Westphal, MD, Founder-CEO of several biotechnology concerns and Managing Partner of Longwood Ventures, Wendy Everett, ScD, of the Network for Excellence in Health Innovation (NEHI, based in Boston), and Nicole Boice, who was founding a new rare disease advocacy organization, Global Genes, to be based in Orange County, California.

During this period, Termeer also pursued investing in young, innovative, biotech start-ups. This gave Termeer another outlet in which to exercise his curiosity and mercantilist instincts. Over the six years after Genzyme, he would take sizable equity interests in a number of emerging, private biotech companies—ProQR, Arrakis Therapeutics, Aura Biosciences, Lysosomal Therapeutics, X4, Artax, and Amylyx among them.

For partnership in this arena, he turned to Dr. Alan Walts, whom Termeer had met on joining Genzyme. Then a young biochemist and doctoral candidate in Bill Roush's lab at MIT and 13 years Henri's younger, Walts would be mentored by Termeer, and the two became close friends. They had worked together for 35 years.

Corporate and not-for-profit board service also occupied a notable proportion of Termeer's overall portfolio of activity in these last six years of his life. In terms of corporate boards, Abiomed, where he had served as a Non-Executive Director for 28 years, was an important commitment. There, he supported its CEO, Michael Minogue, who had joined in 2004. He and Minogue had been through both good times and corporate struggles together, and although Abiomed, a medical devices company, was not one of the rare disease cohort, it was Termeer's first important outside corporate board directorship. Abiomed would occupy a special place in his life as a result.

Although less substantial in terms of years of board service, Termeer's directorships at Moderna and Allergan were also sources of pride to him. These companies, led by CEOs Dr. Stéphane Bancel and David Pyott, respectively, enjoyed great success during his tenure. At Moderna, a prodigious fund-raising machine and a leader in the mRNA therapeutics arena, Termeer worked closely with its Board Chairman, Dr. Noubar Afeyan, also the Managing Partner of Flagship Pioneering.

In January 2014, Termeer joined the Board at Allergan, a world leader in opthalmalogicals, dermatologicals, and medical aesthetics, approximately six weeks before an unexpected hostile takeover struggle with serial acquirer Valeant Pharmaceuticals would commence. Having just lived through the Sanofi–Genzyme process, Termeer's perspectives on fending off a suitor's unwanted approach were fresh and pertinent. He offered wise counsel to David Pyott and his other fellow board members through a disruptive period, which resulted in Allergan's eventual combination with Actavis in March 2015.

Few of Termeer's post-Genzyme relationships proved more important than the one he enjoyed with John Maraganore, PhD. At the time of Termeer's passing, the two men had known each other close to 20 years, dating back to Maraganore's years at Millennium Pharmaceuticals. The two men had first encountered one another in exploring a business development deal.

The conversations went nowhere, as was often the case, but the friendship had been struck, and when in 2002 Dr. Maraganore became CEO of Alnylam Pharmaceuticals, their interactions became more frequent.

Maraganore had not always bought into Termeer's prospects for success. In good humor, he reminisced about his early assessment of Termeer and Genzyme's business proposition, "When I was at Biogen, we'd sort of snicker ... 'What's he doing? How many patients with Gaucher disease are there? What's he going to do with that? How's it going to be a business?'" Later, having a laugh at his own expense, he concluded, "Well, Henri showed us he could do it, and I guess we know the rest of that story."

The two gradually developed a close, informal mentor–mentee relationship that intensified as Alnylam's growth and complexity accelerated. Maraganore was able to learn from Termeer the public policy aspects of orphan drug development and commercialization.

He was effectively tutored by Termeer as he watched his interactions with Senator Kennedy and other public officials at various events in Boston, "Henri was one of the conveners of those events. He was a master at them. He would say to me, 'John, when people question your price and when people question why it has to be this expensive, you should tell them, "Please, come, let me show you why." Don't be defensive, but rather welcome their interrogation, invite them to your company, show them all the things that you do.' Henri's ability to do that so well, and to invite the dialogue, not run away from it, and to welcome people to meet with him and show them ... he was remarkable. I learned from that."

John Maraganore would be among those to succeed Henri Termeer years later as the industry's spokesman, serving as Chairman of BIO from 2017 to 2019. Walking in Termeer's earlier footsteps, he would become the one to convene and lead such gatherings.

By 2017, Termeer no longer stood on the Genzyme platform, but the ingrained fundamentals remained within him. People would wonder where he had the time to support so many. His effectiveness and vision in developing the next generation of leaders to carry forth his mission had been established.

Henri Termeer had already made enormous contributions in fathering the rare disease industry and building a highly successful multinational corporation in Genzyme, but one could argue that this period of Termeer's life was as consequential, and indeed important, as any other. He brought the same intense commitment, drive, focus, and results to aiding scores of leaders who would pursue careers and build companies and organizations in the Genzyme mold. The shadow and importance of Termeer's work in this short six-year period will be seen long into the future.

Days after his sudden, unexpected passing due to a heart attack on May 12, 2017, Henri Termeer's life would be celebrated by many, but few touched people the way his daughter, Adriana, would. Her eulogy, one of six, delivered eight days later on a crisp spring morning at his memorial service held at MIT's Kresge Auditorium, would be remembered by those 600 who witnessed it as one of the most poignant and moving of any they would perhaps ever experience.

It was her story, in her words. A poised, wise-beyond-her-years, 16-year-old daughter would recall her father's reading of a bedtime story to her as a young girl when he had gotten home late one night from the office. It was Shel Silverstein's moving story, *The Giving Tree*, an allegory of how an apple tree would shed its parts, one by one, to help a young boy in need—divesting its fruit, its branches, and ultimately its trunk. Having no further parts to give, the tree had but a stump to offer on which it invited the young boy to sit.

But what it really became, as she revealed each layer of this heartrending story, was a metaphor that evoked memories of her father's unending generosity. In her unwavering, inspiring teenage voice, many heard an echo of Henri Termeer himself.

Packaged in the shell of a tough, but sensitive Dutchman who had the unique mix of conscience and courage to solve a centuries-old problem—the treatment of unfortunate souls born with a genetic disorder—Henri Termeer will be remembered for many things. He will be remembered for the twinkle in his eye, his green laser pointer, his charismatic smile, his stubbornness, and, yes, even Genzyme's high drug prices.

But nothing will be remembered more than his greatest achievement, an achievement that would change the world, that of providing hope to the world's 60 million, forgotten, rare disease patients—people who, in 1983, had no hope and who were destined, in many cases, to die young and painfully.

Much remains to be done in this cause, but Termeer's legion of passionate followers, those who believed in the goodness of his mission and saw his impact and how he achieved it, continues to carry forth his life's work. He fed this legion fish, but he also taught its leaders how to fish on their own.

Henri Termeer's impact has multiplied across the world. He changed the world. And for these things, he will never be forgotten.

Epilogue: Forgotten No More

In the fall of 1985, at the age of 39, Henri Termeer made what became one of the most important, fateful decisions in the history of rare diseases. Contravening the advice of all others and serving the cause of saving a young boy's life, his decision to lead Genzyme's development of Ceredase would usher in one of biotechnology's first struggles in a broader cause for humanity—the research, development, and ultimate treatment of rare diseases.

Henri Termeer's decision would prove to be a seminal moment, numbered among the few that would create, worldwide, an intensity of passion and newfound hope for these patients. Its future impact, at the time, was unimaginable.

Not only would Termeer's decision directly affect the lives of 5,000 Gaucher patients, their families, and their caregivers, but it would also forever change the lives of those living within the global rare disease community—estimated today at half a billion people. In addition, it would, through Genzyme's corporate success and development, show the way to others who would follow.

In the last three and a half decades since the Orphan Drug Act's approval, the world has marveled at the medical advances made in treating many rare disease patients through the application of biotechnology.

When Henri Termeer joined Genzyme, the number of approved products for rare disease indications could be counted on two hands, and these were largely repurposed medicines that had been developed for larger markets and more prevalent diseases.

From the 1983 passage of the Orphan Drug Act through mid-2019, drug developers had submitted more than 7,500 requests seeking the FDA's formal designation of their drug products as orphan drugs. In 1983, there was but one such request; in 2018 alone, the agency received 337.

Today, there are more than 500 FDA-approved orphan drugs covering more than 700 designated orphan rare disease indications. The FDA granted orphan drug marketing approvals for 81 and 91 indications in 2017 and 2018, respectively. The number of approvals in these two years, when combined, represents the largest for any two-year period since 1983.

Annual orphan drug sales in the U.S. market now represent nearly $50 billion or approximately a tenth of the total market for U.S. prescription drugs. One prominent market research firm recently has forecasted that the percentage will reach 20% in the United States by 2024. The comparable figure in 1997 was 4%.

Worldwide, in 2017, the top 20 global pharma-biotech rare disease companies posted, in the aggregate, nearly $130 billion in orphan drug product sales. More than 40 orphan drugs now each exceed $1 billion in global sales.

Better diagnostics for rare diseases have also been developed or are in development. In many cases, these diseases can take years to diagnose; the average is 7.5 years. More rapid diagnosis of rare diseases will lead to more rapid initiation of therapy and better patient outcomes.

The response to the clarion call for rare disease therapies was slow to develop in the 1980s, but grew steadily as families began to see signs of hope for cures or at least substantial improvements in patient quality of life. Patient advocacy organizations now number in the hundreds; so do new family foundations devoted to rare disease awareness and sponsored research.

Corporate pharmaceutical/biotechnology budgets for rare disease research and development have swelled to keep pace with the overall interest and excitement being generated in bringing new, innovative diagnostics and therapies to these patients. One analyst conservatively estimates that global annual corporate expenditures for rare disease R&D now exceed $45 billion. The number of companies developing rare disease therapeutics and diagnostics is well into the hundreds.

Also, legislators and regulators have reorganized the approval procedures for these new therapies and diagnostics, seeking to support the wave of R&D and new products being introduced.

These advances all stem, in part, from that fateful moment, when Henri Termeer made his stand for not only Brian Berman, but for all of those who would follow. When Henri Termeer joined Genzyme, none of this investment, dedication, infrastructure, and hope for the future existed.

There is much left to be done to solve all of the world's rare diseases. There are more than 7,000 of them. But with each new success—stemming from cutting-edge research methods that embrace gene therapy, antisense, RNA intervention or gene editing, to name a few—there is recognition of the progress that has been and continues to be made.

Today, the outlook for the unfortunates who are diagnosed with a rare disease is brighter than ever. These patients are anything but forgotten, and although more than 6,000 rare diseases remain untreated, prodigious efforts are being made to isolate the genetic causes and address many, if not all, of them.

But the cautionary note, if there is one, is the economic reality of the high costs associated with sustaining payment for these chronic therapies. Like a runaway train, they continue to escalate, crowding public and private sector budgets and carving resources from other societal priorities. One of our society's next big issues will be the continued application of its resources in ever-increasing increments to treat these intractable, devastating diseases that directly afflict a small percentage of the world's population.

Henri Termeer, in joining Genzyme in 1983, was confronted with weighing the concerns of rare disease patients on the scale of society's priorities. We can be thankful that his decision was not only swift, but resolute and enduring.

He has left the world in a better place—for the dreams he dreamed and the conscience and courage he applied in pursuing them.

> *The question is whether societies want these very sick patients treated. I think they do.*
>
> —Henri A. Termeer, 1983

Acknowledgments

A journey of this complexity and duration cannot be successful without the kindness and support offered by so many along its path to completion. It is therefore a great pleasure to acknowledge with gratitude those who assisted me in writing *Conscience and Courage*.

Above all, I extend my deepest thanks to Steve Potter, Managing Partner of Odgers Berndtson, LLC, who provided the funding to enable this book's writing and a research effort that encompassed dozens of trips to conduct recorded interviews of more than 130 people. Also lending support from Odgers Berndtson were Mike Kelly, my partner and the global head of the firm's healthcare sector; Heather Campbell; Mabel Setow and Nancy Scarlatta, who hosted me and many of my interviewees in the Boston office of Odgers Berndtson; and Anne Board, who provided her guidance in preparing the author's proposal. Furthermore, my dear friend and senior partner, Michaël Mellink, provided his support and gracious hospitality on my two trips to the Netherlands to interview Henri Termeer's siblings and nearly ten others. In addition, Gilles Gaudefroy, a Partner and head of Odgers' business in France, is acknowledged for his assistance in arranging my visit to Lyon in search of Termeer's connections with Pasteur Merieux. I also thank Alex Thomson, Nick Brill, and Allen Reed, all Partners of Odgers Berndtson, who joined me in conducting various interviews.

Beyond Odgers Berndtson, I acknowledge and thank William Klein, my primary editor, and in some instances co-writer, who provided enormous assistance as we diligently worked our way through the manuscript's preparation. William is a seasoned op-ed writer, and this was his first book, as it was also mine. We struggled together, learned, and laughed together. Thank you, William, for your steadfast support. Your skills and partnership were powerful in their contribution to the book's composition, presentation, and thematic makeup. I also extend thanks to my dear friend, Sandy Costa, who connected us in the first instance.

No one deserves a more profound expression of gratitude than does Jim Geraghty, who was the friend who inspired and exhorted me from the beginning to write this book. I will never forget our conversation in the summer of 2017 as I was in a period of "discernment," trying to decide if I would dedicate the next two years of my life to writing *Conscience and Courage*. Jim's encouragement, as well as that of his wonderful wife, Joan Wood, was indispensable. Jim's friendship throughout the book's production—from assisting me in arranging interviews to assisting me in editing the book's manuscript, confirming factual details, advising on promotional strategies, and generally providing a steady hand in times of challenge—was invaluable. Jim, thank you. Your generosity of time and energy was nothing short of Termeerian.

In getting under way, a number of other people were generous with their time during my period of discernment. Dr. David Meeker, Caren Arnstein, Dr. Charlie Cooney, Zoltan Csimma, Peter Wirth, and Chris Zook provided helpful soundings as I considered the undertaking.

In this regard, I would also like to recognize and express my appreciation to Dr. Carl Schramm, who was the first to review and comment upon my initial author's proposal. Carl is a seven-time author and a thoughtful, learned man. Carl, thank you for your support. Your wise counsel as I worked through the drafting and publication of the book was invaluable

I recognize with gratitude my friends, Philip Goelet, PhD, and his wife, Anette Hoegh Goelet. Philip and Anette were steadfast in their enthusiasm and support for which I am grateful. Philip studied molecular biology at Cambridge under the watchful eye of Nobelist Dr. Sydney Brenner and later as a Helen Hay Whitney Fellow at Columbia University under Eric Kandel, PhD.

To Philip and my friend, Daniel Marshak, PhD, who did his postdoctoral work at Cold Spring Harbor Laboratory (CSHL), I thank them for connecting me with Dr. John Inglis, Publisher and Editor of Cold Spring Harbor Laboratory Press (CSHLP). Philip was particularly instrumental in initiating my outreach to Dr. Inglis and in describing my mission to Dr. Bruce Stillman, President and CEO of CSHL.

To Henri Termeer's widow, Belinda, and her able assistant, Mary Ellen Mucci, I express deep gratitude. To Linda Rubinstein and Lorie Umanita, who devoted hours to preparing Henri Termeer's papers for archival, I do the same. They were enormously helpful as I accessed and worked my way through a trove of documents—around 200 large file boxes, packed with material accumulated over 60 years and with unimaginable detail, all pertaining to Henri Termeer's life—including his pictures, school report cards, correspondence, news clippings, awards, regulatory submissions, legislative documents, speeches, board appointments, and white papers. They were amazingly resourceful and helpful.

And to Belinda, I extend my special thanks for engaging with me as you worked your way through the grieving process in your loss of Henri. To Adriana and Nicholas, I extend the same thanks. I recognize at times it was painful, but I am hopeful that in hindsight the outcome will please you and that it will be proven to have been at least a tiny bit therapeutic.

To Henri Termeer's siblings—Ineke, Marlies, Paul, Bert, and Roel—I extend my gratitude for the November 2018 afternoon and evening we spent together outside Tilburg at Paul's home. Special thanks to Paul and his wife, Mill, for hosting all of us and for the delightful and delicious dinner they provided. And special thanks to all of the siblings for providing the photographs that related to Henri's formative years in the Netherlands. Your remembrances were touching and helpful in understanding Henri's roots.

I would be remiss if I didn't extend thanks to Eric Rayman, Esq., a partner at Miller Korzenik Sommers Rayman, LLP, for his wisdom and legal advice as we entered into various literary contracts and agreements. His experience and perspective were valuable and appreciated.

I would also like to thank my literary agent, Kneerim & Williams, and its cofounder and Partner, John Taylor Williams, and his colleague, Hope Denekamp, who assisted me in placing *Conscience and Courage* with a publisher. They are consummate professionals, a pleasure to work with, and a credit to their industry.

I extend thanks to Michael Hammerschmidt, Senior Advisor at the Science History Institute (SHI), for his assistance in the retrieval of various materials from the institute's archive. Patrick Shea of the SHI was also helpful in this regard.

I also thank Fred McConkey, Senior Managing Director at Bank of America Merrill Lynch and my former investment banking advisor, who was generous in helping me obtain pricing data for Genzyme's stock over the course of the crisis, a period of constant tension for not only the company's patients but also Genzyme's equity holders.

To Monica Higgins, author of *Career Imprints,* a member of the faculty at Harvard University and a fellow Tuckie, I extend my appreciation for the early counsel she provided as I weighed the approach I would take in writing the book. Monica has built a successful academic career, and our time together was helpful as I sought not only to understand her earlier research of Termeer as a "Baxter Boy," but also how to tap into academic resources and intellectual capital.

With regard to pictures, I thank Bill Sibold, CEO of Sanofi Genzyme and his colleagues: Bo Piela, VP & Head of Communications; Jennifer Pereira, Digital Communications Manager; and Lance Webb, Senior Medical Science Liaison—Rare Disease. They were always responsive when I needed help, and their assistance was greatly appreciated.

To Frank Sasinowski, Esq., and his assistant, Josephine Torrente, I express my gratitude for their help in researching the legislative history pertaining to the passage, amendment, and implementation challenges associated with the Orphan Drug Act.

To Paul Kim, Esq., a partner in the Washington office of Foley Hoag LLP, I thank you for your assistance in accessing the archive of Senator Edward M. Kennedy and gathering background on Senator Kennedy's friendship with Henri Termeer.

To Robert Teitelman, I express thanks for his help in the early stages of preparing the author's proposal and in joining me in five of the first interviews I conducted.

To Norman Barton, I express my appreciation for assisting me in reviewing copies of the academic papers I found useful in researching the clinical development of glucocerebrosidase and the evolution of treatments for Gaucher disease and other lysosomal storage disorders. I also thank Norman for the rights to use certain of the pictures that pertained to the pivotal clinical study of glucocerebrosidase.

A select group of individuals helped me fact-check and, in certain instances, edit material found in chapters with which they had firsthand awareness of the events being described. These individuals included Jim Geraghty, Sandra Poole, Ginger More, Peter Wirth, Chris Viehbacher, Victor Dzau, MD, Abbey Meyers, and Uzma Shah, MD.

I also thank Robin Ely, MD, the mother of Brian Berman, who was instrumental in providing and confirming material that appears in the book's Prologue, *Patient One*, as well as Chapter Four, *Mission: Impossible*. In the early years, Dr. Ely was the definition of a warrior mother who fought tirelessly for her children's health and well-being, going on to support other families whose children suffered from rare diseases.

I acknowledge David Pyott for his efforts in exposing me to the story of the Philips NV enterprise, through sending me Fritz Philips' book, as I assessed the impact of World War II on young Henri Termeer's formative years.

In addition Elliott Hillback went to some lengths to make available for my use a videotape he recorded of the March 31, 2011 dinner at the Algonquin Club at which the Genzyme executive team and alumni said their "goodbyes." It was highly useful in understanding the bittersweet, somber mood of the occasion.

Stephen Groft, PharmD, former Director of the Office of Rare Disease Research at NIH, was extremely generous in assisting me with the composition of the Epilogue. Steve's service for more than four decades as a tireless advocate for rare disease research has immeasurably improved the lives of rare disease patients worldwide. Thank you, Steve.

A large group of individuals, too large to thank individually, is that of the more than 130 interviewees who took the time to meet with me for a recorded interview. In many cases, people welcomed me into their homes. Some made the commute to my Boston office, fighting traffic and, in some instances, inclement weather. Their stories and remembrances of Henri Termeer were nothing short of spectacular. The interviews ranged in duration from a half hour up to three hours. Of those on the longer side, I particularly want to thank Henry Blair, Bob Carpenter, Duke Collier, Geoff Cox, Jack Heffernan, Elliott Hillback, Dave McLachlan, David Meeker, Abbey Meyers, Mary Nathan, Greg Phelps, Sandra Poole, Gabe Schmergel, Sandy Smith, Maryze Schoneveld van der Linde, Peter Wirth, and, of course, Jim Geraghty. The background provided by all those interviewed was immensely helpful and added nuance and color to the manuscript. Nearly half worked for Henri Termeer; several sat on his board or the other way around; many were physicians, patients, or patient advocates. They are listed elsewhere in the book. I extend special thanks and owe a debt of gratitude to all of them.

Given the subject matter, many of my interviews took place in Boston or Cambridge, Massachusetts. On many a night over the past two years, I was fed dinner and offered a place to sleep by Whit and Helen Wagner at their magnificent Beacon Hill penthouse apartment. They are generous friends, and I thank them for their support, enduring my frequent visits with a smile, good humor, a glass of appropriately aged port, and a welcome mat.

I also want to thank the staff members of the Cold Spring Harbor Laboratory Press who were helpful as we marched through the steps of publishing *Conscience and Courage*. Among them, Dr. John Inglis, Executive Director and Publisher, was a superb partner in the process. Thank you, John. He was ably supported by Mala Mazzullo, Executive Assistant to the Publisher; Jan Argentine, Director of Editorial Services; Denise Weiss, Book Production Manager and Designer; Wayne Manos, Director of Product Development and Marketing; Kathleen Bubbeo, Production Editor; Linda Sussman, Director of Publication Services; Steve Nussbaum, Finance Director; and Robert Redmond, Marketing Manager. Also, thanks to Tom Adams, CSH Library and Archives, for digitizing the artwork.

I give special thanks at CSHLP to Inez Sialiano, Project Manager, and Carol Brown, Permissions Coordinator. They schooled me in the art of obtaining permissions and consents for quoted material and photographs. We talked nearly daily as we approached the submission deadline, and I appreciated their diligence and valued their experience.

Last but not least, I acknowledge my beloved wife, Anne, and my three children, Sarah, Peter, and Peyton, and their spouses for supporting me through what has been a rewarding, if at times trying, period. Undertakings of this sort require extraordinary focus, and I thank my family for their understanding and support as I have often been away from home, even when I have been home. Throughout this journey, they have unfailingly offered their patience, perspective, and encouragement for which I am grateful. Without them, *Conscience and Courage: How Visionary CEO Henri Termeer Built a Biotech Giant and Pioneered the Rare Disease Industry* could never have been written.

JOHN HAWKINS
Charlottesville, Virginia
April 8, 2019

Timeline

SELECTED MILESTONES

1940	Jacques and Mary Termeer are married on April 2; five weeks later, the Nazis invade the Netherlands; Jacques enlists in the Dutch army, is captured, is imprisoned in Germany, is released, and returns home.
1946	Henri is born on February 28, 1946 to Mary and Jacques Termeer at home, Heuvelstraat 39, Tilburg, the Netherlands —fourth in line of six children and the second of four boys.
1953	The Termeer family moves to a new house, Burgemeester Suijsstraat 10, near Tilburg University.
1958–1961	Termeer undergoes a period of intense devotion to chess; he idolized Max Euwe, Dutch Grandmaster; his school grades suffered.
1964	Termeer graduates St. Oldulphus Lyceum (high school), Tilburg.
1964	Termeer enters the Royal Netherlands Air Force at Breda as a logistician.
1966	Termeer exits the Royal Netherlands Air Force as a 2nd Lieutenant.
1966	Termeer enrolls at the Economische Hogeschool, Erasmus University, Rotterdam, to study economics.
1968	Termeer departs for the United Kingdom to join Norvic, a shoe retailer, to undergo an internship that would presumably lead to a thesis; he becomes a Group Systems Manager, based in East Anglia.
1971	Accompanied by his British girlfriend, Margaret Riches, Termeer departs in August for the United States to attend the University of Virginia Darden School of Business, Charlottesville.

1971	Termeer marries Margaret Riches on October 8, 1971 in Charlottesville.
1973	Termeer graduates from the University of Virginia Darden School of Business with his Masters in Business Administration.
1973	Termeer joins Baxter Travenol as Assistant to the Vice President of International Marketing, based in Chicago, Illinois.
1974	Termeer becomes International Product Planning Manager, Hyland Therapeutics Division, Baxter, based in Orange County, California.
1974	Termeer is chosen to undertake a special three-month assignment, reporting to Bill Graham, Baxter CEO, based in Brussels.
1974–1976	Termeer becomes International Marketing Manager, Artificial Organs Division, Baxter, based in Chicago.
1976–1979	Termeer becomes General Manager, Baxter Deutschland GmbH, Munich, Germany.
1979–1983	Termeer is appointed Executive Vice President, Hyland Therapeutics, Baxter, based in Glendale, California.
1981	Founded by scientist Henry Blair, entrepreneur Sheridan Snyder, and venture capitalists at Oak Investment Partners, Genzyme is incorporated in Delaware.
1983	The Orphan Drug Act is signed into law by President Ronald Reagan.
1983	In April, through the initiative of cofounder Sheridan Snyder, Genzyme enters into a 10-year agreement with BioInformation Associates (BIA), a corporation owned by a group of eight scientists in the life sciences fields who were full-time faculty members of MIT and Harvard.

1983	In October, Termeer joins Genzyme Corporation, Boston, Massachusetts as the company's President & Director, reporting to cofounder & CEO, Sheridan Snyder. On joining, Termeer's base salary is $110,000, and he is allowed to purchase 108,500 shares of Genzyme common stock for $4.32 per share.
1984	On December 15, Brian Berman becomes the first Gaucher disease patient treated with enzyme replacement therapy and is dosed with human placental glucocerebrosidase.
1985	After having, in April, initiated a search to find his replacement, CEO Snyder threatens to fire Termeer at a December Genzyme Board meeting. At this meeting, the Board chooses a different course, elevating Termeer to become its President and CEO.
1986	Genzyme Corporation goes public on June 6 through a $28.2 million IPO underwritten by Kidder Peabody, Montgomery Securities, and Cowen, valuing the business (pre–money raised) at $53 million.
1988	Termeer is elevated to Chairman, President & CEO, Genzyme Corporation.
1989	Genzyme acquires Integrated Genetics, a critical step in building its mammalian cell culture production capability.
1989	Termeer's marriage to Margaret Riches dissolves in divorce; she returns to live in her native United Kingdom.
1989	Termeer joins the Board of Directors of Abiomed, his first corporate board appointment, and serves 28 years till his passing in 2017.
1991	Ceredase is approved by the U.S. FDA for the treatment of Gaucher disease, a disorder estimated to affect only 5,000 patients worldwide.
1993	William Jefferson Clinton is inaugurated as 43rd U.S. President; the Health Security Act is proposed; Termeer mobilizes and co-leads biotech's response in opposition.

1994	Cerezyme, an rDNA successor product to Ceredase, is approved by the U.S. FDA for the treatment of Gaucher disease.
1995	The Allston Landing manufacturing site is officially opened.
1996	Senator Edward M. Kennedy assumes his mantle as the "Lion of the Senate," projecting profound influence for the next decade over U.S. health legislative affairs; Termeer's friendship with Senator Kennedy deepens.
1997	Jacques Termeer, Henri's father, dies at age 87.
1998	Termeer marries Belinda Herrera, a native of New Mexico, on August 1 at the Termeer summer residence in Biddeford Pool, Maine. They purchase a house in Marblehead, Massachusetts into which they move in May 2000.
1999	Termeer becomes a naturalized U.S. citizen; the certificate is issued on October 20, 1999.
2003	In April, Genzyme's rare disease therapies Fabrazyme and Aldurazyme are both approved by the U.S. FDA.
2004	Termeer joins the Board of Massachusetts General Hospital, a position he held till his death.
2006	Genzyme's rare disease therapy Myozyme is approved by the U.S. FDA.
2006–2017	Termeer joins the Board of Members of MIT Corporation in 2006; he is later selected in 2011 to serve on its Executive Committee, a position he held until his death.
2007	Termeer joins the Board of the Federal Reserve Bank of Boston as its Deputy Chairman.
2008	The Termeers acquire the Yellow House, a stately property overlooking the Marblehead Harbor, which would later serve as Termeer's office (2011–2017).
2009	The crisis at Allston Landing begins as the Vesivirus 2117 contamination is announced on June 16 and the plant is shut down to be decontaminated; product outages and rationing ensue. Patients and their physicians are infuriated.
2010–2011	Termeer serves two full years as Chairman of the Federal Reserve Bank of Boston.

2010	Activist investors Carl Icahn and Ralph Whitworth accumulate large shareholdings in Genzyme, assert their power, and designate four directors to serve on Genzyme's Board of Directors.
2010	Based on 2009 revenues of $4.58 billion, Genzyme is named for the first time as a Fortune 500 company.
2010	A Consent Decree of Permanent Injunction is filed in May against Termeer and Genzyme in connection with the operation of its Allston Landing facility. Sanofi Aventis approaches Genzyme to commence merger discussions.
2011	On February 16, 2011, Sanofi and Genzyme jointly announce Sanofi would acquire Genzyme for $20.1 billion in cash plus additional potential CVR consideration of $3.0 billion. The combination is declared a "New Beginning."
2011	On April 8, 2011, Sanofi's acquisition of Genzyme closes. Genzyme's shares are delisted and the company is merged into Sanofi, becomes a new division, and is rebranded Sanofi Genzyme. David Meeker, MD, assumes leadership of Sanofi Genzyme as its CEO and successor to Henri Termeer.
2011–2017	Termeer begins a six-year period of mentoring scores of leaders, accepting and fulfilling the duties of many board directorships, backing life sciences entrepreneurs, investing in young promising biotechnology enterprises, and serving others—in the corporate as well as not-for-profit arenas.
2012	Made possible through the Termeer's generous $10 million gift, the Henri and Belinda Termeer Center for Targeted Therapies is dedicated and opened at Massachusetts General Hospital.
2014	Mary Termeer, Henri's mother, passes away at age 100.
2017	On Friday, May 12, 2017, at the age of 71, Henri Termeer dies in Marblehead, Massachusetts. A funeral mass is held at Our Lady Star of the Sea Church, Marblehead on May 22, 2012, after which he is laid to rest at nearby Waterside Cemetery.

NOT-FOR-PROFIT BOARD APPOINTMENTS

1995–2017	Biomedical Science Careers Programs
1987–2011	Biotechnology Innovation Organization (BIO) Vice Chairman, 1993–1995; Chairman, 1995–1997; Chairman Emeritus, 2012–2017
2008–2017	Boston Ballet—Board of Trustees
2009	Commonwealth of Massachusetts—Governor's Board of Economic Advisors
2007–2011	Federal Reserve Bank of Boston Deputy Chairman, 2007–2009; Chairman, 2010–2011
2000–2017	Harvard Medical School Board of Fellows
2011–2017	Harvard Medical School—Therapeutics Advisory Council
2004–2017	Massachusetts General Hospital
2006–2017	Massachusetts Institute of Technology (MIT) Board of Members
2011–2017	MIT—Executive Committee, Board of Members
1992–2017	Museum of Science, Boston
2002–2017	Network for Excellence in Health Innovation (NEHI)
2008–2017	Partners HealthCare
2002–2011	Pharmaceutical Research and Manufacturers Association (PhRMA)
2006–2017	Project Hope
2013	WGBH

CORPORATE BOARD APPOINTMENTS

1987–2017	Abiomed
2014–2015	Allergan

2011–2017	Aura Biosciences
1992–2003	Autoimmune
2011–2016	Aveo
1996–2002	Diacrin
1996–2002	Dyax
Appointed 1992	Geltex
1983–2011	Genzyme
1998–2002	Genzyme Transgenics IG Laboratories Introgene
Appointed 1993	Lotus Development
2005–2015	Medical Simulation
2013–2017	Moderna Therapeutics Neozyme II
2014–2017	ProQR Therapeutics
2011–2016	Verastem Xenova

SELECTED INDUSTRY AWARDS AND HONORS

2008	BIO-Science History Institute, Biotechnology Heritage Award
2010	Boston Biotech, CEO Lifetime Achievement Award
2007	Ernst & Young, Master Entrepreneur Award
2009	Frost & Sullivan, Pharmaceuticals & Biotechnology Lifetime Achievement Award
1991, 1994	Laguna Niguel Conference, The Best of Biotech Award

1997	Laguna Niguel Conference, Hall of Fame
1992	Merrill Lynch, Ernst & Young, Entrepreneur of the Year
1995	*SUCCESS* Magazine, Renegade of the Year
1990, 1991, 1992	*The Wall Street Transcript,* CEO Gold Award

SELECTED ACADEMIC AWARDS AND HONORS

1999	American Academy of Arts & Sciences, Inducted—Fellow
2010	Babson College, Distinguished Academy of Entrepreneurs
2005	British Royal College of Physicians, Inducted—Honorary Fellow
2011	Framingham State University, Honorary Degree, Doctor of Science
2011	Northeastern University, Honorary Degree, Doctor of Global Business
2007	University of Massachusetts, Honorary Degree, Doctor of Science
2011	University of Twente (Netherlands), Honorary Degree, Doctorate
1996	Worcester State College, Honoris Causa, Doctor of Business Administration

SELECTED COMMUNITY AWARDS AND HONORS

1995	Anti-Defamation League, Torch of Liberty Award
1997	Commonwealth of Massachusetts, Governor's New American Award
1997	Cardinal Cushing School for Exceptional Children, Humanitarian of the Year

1999	Biomedical Science Careers Program, The Hope Award
1999	International Institute of New England, The Golden Door Award
2000	Genetic Disease Foundation, Humanitarian Award
2003	American Heart Association, Cor Vitae Award
2003	March of Dimes, Franklin Delano Roosevelt Humanitarian Award
2005	United States of America, National Medal of Technology and Innovation (accepted on behalf of Genzyme)
2010	Netherlands American Foundation, Ambassador William J. Middendorf Award
2012	National Tay–Sachs & Allied Diseases, Gala Honoree
2012	Global Genes, Project RARE Lifetime Achievement Award
2015	World Orphan Drug Congress, USA Lifetime Achievement Award

Sources

The interviews (held October 20, 2017 through February 5, 2019) provided personal accounts utilized in providing context and factual history for each chapter, along with the primary research, references, and additional readings. Direct quotes are keyed to page numbers. A master interviewee list follows.

Prologue: Patient One

Interviews and Background Information

Barton, Norman
Boice, Nicole
Drake, Peter
Ely, Robin A., *see* pages 2–4
Haffner, Marlene
Lodish, Harvey
Meyers, Abbey
More, Eileen (Ginger)
Moscicki, Richard, *see* page 5
Papadapoulos, Stelios
Rubinstein, Linda
Sherblom, Jim
Taunton-Rigby, Alison
Zerhouni, Elias

Henri A. Termeer Quotes and Sources

Page 2, ¶3. Clarke, Toni. Special report: inside the battle for Genzyme's future. *Reuters*, news story, May 5, 2010. https://www.reuters.com/article/us-usa-biotech-genzyme-idUSTRE6445JA20100505

Page 4, last ¶. Panel remarks, Biotech Showcase, January 2017. Reported by Richetti E. 2017. The advocates of rare disease advocates: remembering Henri Termeer. *Partnering Insight*, July 15, 2017. https://knect365.com/partnering-insight/article/c333f395-beca-406d-8e2d-e3f48a9670a9/the-advocate-of-rare-disease-advocates-remembering-henri-termeer

Page 5, ¶10. Usdin S. 2017. Henri: *N* of 1, how the late Genzyme CEO Henri Termeer created the orphan drug industry. *BioCentury* May 19, 2017. https://www.biocentury.com/biocentury/strategy/2017-05-19/how-late-enzyme-ceo-henri-termeer-created-orphan-drug-industry

Page 6, ¶7. Crawford J. 2010. The business of saving lives. *Babson Magazine*, Spring 2010.

References/Additional Readings

Trams EG, Brady RO. 1960. Cerebroside synthesis in Gaucher's disease. *J Clin Invest* **39:** 1546–1550.

Brady RO, Kanfer J, Shapiro D. 1965. The metabolism of glucocerebrosides. I. Purification and properties of a glucocerebroside-cleaving enzyme from spleen tissue. *J Biol Chem* **240:** 39–43.

Brady RO, Kanfer J, Shapiro D. 1965. Metabolism of glucocerebrosides. II. Evidence of an enzymatic deficiency in Gaucher's disease. *Biochem Biophys Res Commun* **18:** 221–225.

Brady RO, Kanfer JN, Bradley RM, Shapiro D. 1966. Demonstration of a deficiency of glucocerebroside-cleaving enzyme in Gaucher's disease. *J Clin Invest* **45:** 1112–1115.

Brady RO. 1966. The sphingolipidoses. *N Engl J Med* **275:** 312–318.

Brady RO, Kanfer JN, Mock MB, Fredrickson DS. 1966. The metabolism of sphingomyelin. II. Evidence of an enzymatic deficiency in Niemann–Pick disease. *Proc Natl Acad Sci* **55:** 366–369.

Brady RO, Gal AE, Bradley RM, Martensson E, Warshaw AL, Laster L. 1967. Enzymatic defect in Fabry's disease. Ceramidetrihexosidase deficiency. *N Engl J Med* **276:** 1163–1167.

Kampine JP, Brady RO, Kanfer JN, Feld M, Shapiro D. 1967. Diagnosis of Gaucher's disease and Niemann–Pick disease with small samples of venous blood. *Science* **155:** 86–88.

Brady RO. 1969. Tay–Sachs disease. *N Engl J Med* **281:** 1243–1244.

Kolodny EH, Brady RO, Volk BW. 1969. Demonstration of an alternation of ganglioside metabolism in Tay–Sachs disease. *Biochem Biophys Res Commun* **37:** 526–531.

Sloan HR, Uhlendorf BW, Kanfer JN, Brady RO, Fredrickson DS. 1969. Deficiency of sphingomyelin-cleaving enzyme activity in tissue cultures derived from patients with Niemann–Pick disease. *Biochem Biophys Res Commun* **34:** 582–588.

Brady RO, Johnson WG, Uhlendorf BW. 1971. Identification of heterozygous carriers of lipid storage diseases. Currrent status and clinical applications. *Am J Med* **51:** 423–431.

Brady RO, Uhlendorf BW, Johnson WG. 1971. Fabry's disease: antenatal detection. *Science* **172:** 174–175.

Epstein CJ, Brady RO, Schneider EL, Bradley RM, Shapiro D. 1971. In utero diagnosis of Niemann–Pick disease. *Am J Hum Genet* **23:** 533–555.

Ho MW, Seck J, Schmidt D, Veath ML, Johnson W, Brado RO, O'Brien JS. 1972. Adult Gaucher's disease: kindred studies and demonstration of deficiency of acid β-glucosidase in cultured fibroblasts. *Am J Hum Genet* **24:** 37–45.

Johnson WG, Brady RO. 1972. Ceramide trihexosidase from human placenta. *Methods Enzymol* **28:** 849–856.

Schneider RO, Ellis WG, Brady RO, McCulloch JR, Epstein CJ. 1972. Infantile (type II) Gaucher's disease: in utero diagnosis and fetal pathology. *J Pediatrics* **81:** 1134–1139.

Tallman JF, Johnson WG, Brady RO. 1972. The metabolism of Tay–Sachs ganglioside: catabolic studies with lysosomal enzymes from normal and Tay–Sachs brain tissue. *J Clin Invest* **51:** 2339–2345.

Johnson WG, Desnick RJ, Long DM, Sharp HL, Krivit W, Brady B, Brady RO. 1973. Intravenous injection of purified hexosaminidase A into a patient with Tay–Sachs disease. *Birth Defects Orig Artic Ser* **9:** 120–124.

Brady RO, Tallman JF, Johnson WG, Gal AE, Leahy WR, Quirk JM, Dekaban AS. 1973. Replacement therapy for inherited enzyme deficiency. Use of purified ceramide trihexosidase in Fabry's disease. *N Engl J Med* **289:** 9–14.

Pentchev PG, Brady RO, Hibbert SR, Gal AE, Shapiro D. 1973. Isolation and characterization of glucocerebrosidase from human placental tissue. *J Biol Chem* **248:** 5256–5261.

Brady RO, Pentchev PG, Gal AE, Hibbert SR, Dekaban AS. 1974. Replacement therapy for inherited enzyme deficiency. Use of purified glucocerebrosidase in Gaucher's disease. *N Engl J Med* **291:** 989–993.

Pentchev PG, Brady RO, Gal AE, Hibbert SR. 1975. Replacement therapy for inherited enzyme deficiency. Sustained clearance of accumulated glucocerebroside in Gaucher's disease following infusion of purified glucocerebrosidase. *J Mol Med* **1:** 73–78.

Beutler E, Dale GL, Guinto DE, Kuhl W. 1977. Enzyme replacement therapy in Gaucher's disease: preliminary clinical trial of a new enzyme preparation. *Proc Natl Acad Sci* **74:** 4620–4623.

Furbish FS, Blair HE, Shiloach J, Pentchev PG, Brady RO. 1977. Enzyme replacement therapy in Gaucher's disease: large-scale purification of glucocerebrosidase suitable for human administration. *Proc Natl Acad Sci* **74:** 3560–3563.

Brady RO, Furbish FS. 1982. Enzyme replacement therapy: specific targeting of exogenous enzymes to storage cells. In *Membranes and transport,* Vol. 2 (ed. Martonosi AN) , pp. 587–592. Plenum, New York.

Barton NW, Furbish FS, Murray GJ, Garfield M, Brady RO. 1990. Therapeutic response to intravenous infusions of glucocerebrosidase in a patient with Gaucher disease. *Proc Natl Acad Sci* **87:** 1913–1916.

Barton NW, Brady RO, Dambrosia JM, Di Bisceglie AM, Doppelt SH, Hill SC, Mankin HJ, Murray GJ, Parker RI, Argoff CE et al. 1991. Replacement therapy for inherited enzyme deficiency—macrophage-targeted glucocerebrosidase for Gaucher's disease. *N Engl J Med* **324:** 1464–1470.

Murray GJ, Howard KD, Richards SM, Barton NW, Brady RO. 1991. Gaucher's disease: lack of antibody response in 12 patients following repeated intravenous infusions of mannose terminal glucocerebrosidase. *J Immunol Methods* **137:** 113–120.

Grabowski GA, Barton NW, Pastores G, Dambrosia JM, Banerjee TK, McKee MA, Parker C, Schiffmann R, Hill SC, Brady RO. 1995. Enzyme therapy in type 1 Gaucher disease: comparative efficacy of mannose-terminated glucocerebrosidase from natural and recombinant sources. *Ann Internal Med* **122:** 33–39.

General Background Sources

Conversations with Henri Termeer, oral history conducted by Ted Everson, Jennifer Dionisio, Pei Koay, and Arnold Thackray, May 23, December 7, 2006; August 2, 2007; December 18, 2008; and September 30, 2011; edited by Gavin Rynne and Mark Jones. © 2012 The Life Sciences Foundation, San Francisco.

Mikami K. 2017. Orphans in the market: The history of Orphan Drug policy. In *Social history of medicine,* Vol. 1, pp. 1–22, hkx098. Oxford University Press, Oxford. https://doi.org/10.1093/shm/hkx09

Chapter 1: The Leader Within

Interviews and Background Information

Robin, Ely
Littlechild, John
Taminiau (née Termeer), Ineke, *see* page 13
Termeer, Belinda
Termeer, Bert, *see* pages 11–12
Termeer, Nicholas
Termeer, Paul, *see* page 12
Termeer, Roel, *see* page 15
Verduijn (née Termeer), Marlies, *see* page 10

Henri A. Termeer Quotes and Sources

Pages 8–10, 12, 13 ¶7, page 14 ¶5 and ¶8, page 15 ¶5, page 16 ¶1 and ¶3. *Conversations with Henri Termeer,* oral history conducted by Ted Everson, Jennifer Dionisio, Pei Koay, and Arnold Thackray, May 23, December 7, 2006; August 2, 2007; December 18, 2008; and September 30, 2011; edited by Gavin Rynne and Mark Jones. © 2012 The Life Sciences Foundation, San Francisco.

Page 13, ¶2. Belinda Termeer interview.

Page 14, ¶1–3. McPherson C. 1997. "Henri Termeer: an oral history." High school student paper, November 17, 1997 (family collection).

General Background Source

Philips F. 1976. *45 years with Philips: an industrialist's life.* Blandford, Poole, United Kingdom

Chapter 2: Wings

Interviews and Background Information

Barrell, Alan, *see* page 25
Casamento, Chuck
Castaldi, Dave, *see* page 29
Carpenter, Bob, *see* pages 27, 29
Chistensen, Gustav
Cooney, Charles
Drake, Peter
Ely, Robin
Gantz, Bill, *see* page 28
Geraghty, James
Hillback, Elliott
Huyett, Bill
Littlechild, John
More, Eileen (Ginger)
Moufflet, Gerard
Papadopoulos, Stelios
Phelps, Greg
Schmergel, Gabriel, *see* page 24
Smith, C. Ray
van Heek, Jan

References and Additional Readings

The University of Chicago News Office Press Release. 1997. William B. and Catherine V. Graham give $10 million to the University of Chicago Center for Continuing Studies renamed in their honor. University of Chicago, Illinois, March 18.
Higgins MC. 2005. *Career imprints: creating leaders across an industry.* Jossey-Bass, San Francisco.

Chapter 3: A Great Convergence

Interviews and Background Information

Aldrich, Richard
Blair, Henry, *see* pages 32, 40
Boger, Joshua
Carpenter, Robert
Casamento, Chuck
Castaldi, Dave
Cooney, Charlie, *see* page 39
Cox, Geoffrey
Ely, Robin
Fleming, Dave
Gantz, Bill
Geraghty, James, *see* pages 37–38
Greene, Ted, *see* page 34
Haffner, Marlene
Heffernan, Jack
Hillback, Elliot
Huyett, Bill
Littlechild, John, *see* pages 36, 40
Lodish, Harvey
Meeker, David
Meyers, Abbey, *see* pages 41–42
More, Eileen (Ginger), *see* pages 32–33
Phelps, Greg
Pops, Richard
Schmergel, Gabe, *see* page 37
Sharp, Phil
Sherblom, Jim
van Heek, Jan
Walts, Alan
Wirth, Peter
Zerhouni, Elias

References and Additional Readings

Green H. 2008. Interview conducted by Matthew Shindell, October 8. The San Diego Technology archive (SDTA), UC San Diego Library, La Jolla, CA, with permission.

Steele J. 2008. 30th Hybritech reunion marks biotech's genesis. *San Diego Union Tribune*, September 14.

Golden F. 1981. Shaping life in the lab: the boom in genetic engineering. *Time*, March 9.

Watson N. 2003. This Dutchman is flying maverick. Biotech firm Genzyme is winning big profits from a contrarian strategy: think small. *Fortune*, June 23.

Waxman HA. 1986. The history and development of the Orphan Drug Act. In *Orphan diseases and orphan drugs* (eds. Scheinberg IH, Walshe JM), pp. 135–149. Manchester University Press, Manchester, NH.

General Background Sources

Padgett JF, Powell WW. 2012. *The emergence of organizations and markets*, pp. 379–433. Princeton University Press, Princeton, NJ.

Papadopoulos S. 2003. Going public with a valuation higher than $300 million may now be necessary to attract interest from institutional investors. *Nature* doi:10.1038/89401. https://www.nature.com/bioent/2003/030101/full/nbt0601supp_BE18.html

Meyers AS. 2016. *Orphan drugs: a global crusade.* www.abbeysmeyers.com

Chapter 4: Mission: Impossible

Interviews and Background Information

Barton, Norman
Berger, Jacques
Blair, Henry, *see* pages 46–47
Carpenter, Robert
Casamento, Chuck
Cooney, Charles
Cox, Geoffrey, *see* page 54
Drake, Peter, *see* page 49
Ely, Robin A, *see* page 55
Fleming, Dave
Furbish, Scott, *see* page 51
Geraghty, James
Haffner, Marlene
Heffernan, Jack
Hillback, Elliott
Hockfield, Susan
Lawton, Alison
Lodish, Harvey
Maraganore, John
Meeker, David
Meyers, Abbey
Mistry, Pramod
More, Eileen (Ginger), *see* pages 45–46
Moscicki, Richard
Nathan, Mary
Papadopoulos, Stelios
Phelps, Greg
Pops, Richard
Rubinstein, Linda
Schmergel, Gabe
Sharp, Phil
Sherblom, Jim, *see* pages 45–47, 53–55
Taunton-Rigby, Alison, *see* pages 53–54
Tayot, Jean-Louis
van Heek, Jan
Walts, Alan
Webb, Lance
Wirth, Peter, *see* pages 47–48

References and Additional Readings

Bartlett CA. 2002. *Genzyme's Gaucher initiative: global risk and responsibility.* Harvard Business School, Boston, 5-303-066, December 13.

Bartlett CA, McLean AN. 2002. *Genzyme's Gaucher initiative: global risk and responsibility.* Harvard Business School, N9-303-048, September 10.

Chapter 5: Delivering Hope

Interviews and Background Information

Aliski, William
Barton, Norman, *see* pages 58–59
Boice, Nicole
Buyers, Rhonda, *see* pages 62–63
Coolidge, Kathleen
Cooney, Charles
Cox, Geoffrey
Crowley, John
Daley, George
Drake, Peter
Ely, Robin
Feldbaum, Carl, *see* pages 61–62
Geraghty, James
Haber, Dan
Haffner, Marlene
Heffernan, Jack, *see* page 57
Hillback, Elliott
Howe, John
Johnson, Jack
Kaye, Ed
King, John
Lawton, Alison
Leonard, Dan
Lodish, Harvey
Maraganore, John
Meeker, David, *see* page 60
Merrifield, Ann
Meyers, Abbey, *see* page 59
Mistry, Pramod
Moscicki, Richard, *see* page 65
Nathan, Mary
Papadapoulos, Stelios
Piela, Bo
Ring, Jamie
Rubinstein, Linda
Schoneveld van der Linde, Maryze
Secor, Alicia
Shah, Uzma
Sherblom, Jim
Skaletsky, Mark
Smith, Alan
Taunton-Rigby, Alison, *see* page 58
Terry, Sharon
Thier, Samuel
Tierney, Tomye, *see* page 62
van Heek, Jan
Walts, Alan
Webb, Lance
Zerhouni, Elias

References and Additional Readings

Reilly PR. 2015. *Orphan: the quest to save children with rare genetic disorders.* Cold Spring Harbor Laboratory Press, Cold Spring Harbor, NY.

General Background Sources

Deegan PB, Cox TM. 2012. Imiglucerase in the treatment of Gaucher disease: a history and perspective. *Drug Des Devel Ther* **6:** 81–106.

Pollack A. 2000. Two paths to the same protein. *The New York Times,* March 28.

Chapter 6: Duty, Honor, Patients

Interviews and Background Information

Aliski, William
Boice, Nicole
Buyers, Rhonda, *see* pages 70–71
Coolidge, Kathleen, *see* page 71
Coughlin, Bob
Cox, Geoffrey
Crowley, John
Daley, George
Drake, Peter, *see* page 68
Dzau, Victor
Feldbaum, Carl
Geraghty, James
Haber, Dan
Haffner, Marlene, *see* page 75
Heffernan, Jack
Hillback, Elliott
Howe, John
Johnson, Jack
Kaye, Ed
King, John, *see* page 75
Lawton, Alison
Leonard, Dan
Maderis, Gail
Meeker, David
Merrifield, Ann
Meyers, Abbey, *see* pages 76–77
Mistry, Pramod
Moscicki, Richard
Nathan, Mary
Piela, Bo
Poole, Sandra, *see* pages 72–73
Pops, Richard, *see* page 76
Ring, Jamie, *see* page 72
Rubinstein, Linda
Saltonstall, Peter
Schoneveld van der Linde, Maryze, *see* pages 73–74
Secor, Alicia
Shah, Uzma
Sherblom, Jim
Skaletsky, Mark
Smith, Alan
Smith, Sandy
Taunton-Rigby, Alison
Terry, Sharon
Thier, Samuel
Tierney, Tomye
van Heek, Jan
Webb, Lance, *see* pages 69–71
Zerhouni, Elias

General Background Source

Lincoln E. 2017. *Patient-centricity: answering industry's key questions.* LifeScienceLeader.com, pp. 42–43, July 2017.

Chapter 7: Into the Lion's Den

Interviews and Background Information

Aliski, Bill, *see* pages 85–86
Buyers, Rhonda
Carpenter, Robert
Coughlin, Bob

Collier, Duke, *see* pages 81–82
Crowley, John
Drake, Peter
Dunsire, Deborah
Dzau, Victor
Feldbaum, Carl
Froehlich, Sara
Geraghty, James
Greenwood, Jim, *see* page 93
Haffner, Marlene
Hillback, Elliott
Holcombe, Kay
Lawton, Alison
Mack, Connie
Maderis, Gail
McGrane, Mary
McLachlan, David
Meyers, Abbey, *see* pages 84, 90
Mistry, Pramod
Moscicki, Richard
Myers, Michael, *see* pages 91–92
Nathan, Mary
Papadopoulos, Stelios
Phelps, Greg
Pops, Richard
Rosenblatt, Mike
Sasinowski, Frank
Schmergel, Gabe
Smith, Alan
Smith, Sandy
van Heek, Jan
Webb, Lance
Wirth, Peter

References and Additional Readings

Beutler E. 1991. Gaucher's disease. *N Engl J Med* **325:** 1354–1360.

Cushman JH Jr. 1992. Incentives for research on drugs are debated. *The New York Times,* January 22. https://nyti.ms/29aTy4a

Figueroa ML, Rosenbloom BE, Kay AC, Garver P, Thurston DW, Koziol JA, Gelbart T, Beutler E. 1992. A less costly regimen of alglucerase to treat Gaucher's disease. *N Engl J Med* **327:** 1632–1636.

Garber AM. 1992. No price too high. *N Engl J Med* **327:** 1676–1678.

Garber AM, Clark AE, Goldman DP, Gluck ME. 1992. Federal and private roles in the development and provision of alglucerase therapy for Gaucher disease. U.S. Congress Office of Technology Assessment, Report No. OTA-BP-H-104, Washington, DC.

Kroll J. 2008. Howard's end: Metzenbaum was true to form through his last days in the Senate. *Plain Dealer Extra,* March 12 (originally published in *The Plain Dealer* December 4, 1994). jkkroll@plaind.com.

Moscicki RA, Taunton-Rigby A. 1993. Treatment of Gaucher's disease. *N Engl J Med* **328:** 1564; author reply, 1567–1568.

General Background Sources

Metzenbaum HM. 1992. Anticompetitive abuse of the Orphan Drug Act: Invitation to high prices. Hearing before the subcommittee on antitrust, monopolies and business rights of the Committee on the Judiciary U.S. Senate, January 21, 1992, Serial No. J-102-48, Washington, DC.

Termeer H. 1993. The cost of miracles. *The Wall Street Journal,* op-ed, November 16.

Rosenblatt M, Termeer H. 2017. Reframing the conversation on drug pricing. *N Engl J Med (Catalyst).* https://catalyst.nejm.org/reframing-conversation-drug-pricing/

Augustine NR, Madhaven G, Nass SJ. 2018. *Making medicines affordable: a national imperative,* 1st ed. National Academies Press, Washington, DC.

Chapter 8: From Brazil to China

Interviews and Background Information

Cox, Geoffrey
Crowley, John
Csimma, Zoltan
Daley, George
Dunsire, Deborah
Feldbaum, Carl
Fleming, Dave
Geraghty, James, *see* page 106
Haffner, Marlene
Heffernan, Jack
Hillback, Elliott
Incerti, Carlo
Howe, John, *see* pages 102–103
Kaye, Ed
Lawton, Alison
Mack, Connie
Maderis, Gail
McDonough, Geoff
McLachlan, David
Meeker, David
Meijer, Dick, *see* pages 105–106
Merrifield, Ann
Mistry, Pramod, *see* pages 102–103
Moscicki, Richard
Piela, Bo
Poole, Sandra
Ring, Jamie
Rosenstein, Linda
Schoneveld van der Linde, Maryze
Secor, Alicia
Shah, Uzma, *see* pages 95–98
Smith, Alan
Smith, Sandy, *see* pages 98, 100
Taunton-Rigby, Alison
Tierney, Tomye, *see* pages 96–97
van Heek, Jan
Vivaldi, Rogerio, *see* pages 98–101
Webb, Lance
Xue, James, *see* pages 103–104
Zerhouni, Elias

General Background Sources

Bartlett CA, McLean AN. 2002. *Genzyme's Gaucher initiative: global risk and responsibility.* Harvard Business School, Boston.

Project HOPE Report. 2005. *South Asia: Genzyme and Project HOPE partner to rebuild critical health care programs in tsunami-affected areas.* Project HOPE Report, published July 13, https://reliefweb.int

Chapter 9: HAT Tips

Interviews and Background Information

Albers, Jeff
Arnstein, Caren
Butler, John
Carpenter, Robert
Collier, Duke
Cox, Geoffrey
Csimma, Zoltan
Dzau, Victor, *see* page 111
Enyedy, Mark
Feldbaum, Carl
Geraghty, James
Heffernan, Jack, *see* page 111
Hillback, Elliott
Huyett, Bill
Incerto, Carlo
Kaye, Ed, *see* pages 112–113

Lawton, Alison, *see* pages 109–110
Maderis, Gail
McDonough, Geoff
McLaughlan, David
Merrifield, Ann
Meyers, Abbey
Moscicki, Richard
Phelps, Greg, *see* page 107
Piela, Bo, *see* page 114
Poole, Sandra
Ring, Jamie
Secor, Alicia, *see* page 109
Smith, Sandy, *see* pages 107–108
Soteropoulos, Paula
Tierney, Tomye
van Heek, Jan, *see* page 108
Vivaldi, Rogerio
Walts, Alan
Webb, Lance
Wirth, Peter, see page 110
Wood, Joan, *see* page 110
Xue, James

General Background Source

Termeer HA. 2014. A biotechnology entrepreneur's story: advice to future entrepreneurs. In *Biotechnology entrepreneurship*, pp. 15–20. Elsevier, New York.

Chapter 10: The Crisis

Interviews and Background Information

Arnstein, Caren, *see* page 120
Bamforth, Mark
Carpenter, Robert
Coughlin, Bob
Csimma, Zoltan
Drake, Peter
Dunshire, Deborah
Dzau, Victor
Feldbaum, Carl
Ferrer, Carlos
Geraghty, James
Heffernan, Jack
Hillback, Elliott
Huyett, Bill
Johnson, Jack, *see* page 122
Kaye, Ed, *see* page 121
Mack, Connie
Maderis, Gail
Maraganore, John
McDonough, Geoff, *see* page 121
Meeker, David
Merrifield, Ann
Moscicki, Richard
Okita, Blair
Papadopoulos, Stelios, *see* page 117
Piela, Bo
Poole, Sandra, *see* pages 119–120, 122
Pops, Richard
Secor, Alicia
Sharp, Phil
Skaletsky, Mark
Smith, Sandy
Viehbacher, Chris
Walts, Alan
Webb, Lance
Wirth, Peter
Wood, Joan
Zerhouni, Elias

References and Additional Readings

Lord R. 2011. Patients suffer as drug maker rations Fabrazyme medicine. *Pittsburgh Post-Gazette*, July 10.

McDonough G. 2009. Genzyme provides update on Cerezyme supply and 2009 financial information. Sanofi Genzyme News release, August 10, EX-99.1 2 a09-16351_4ex99d1.htm EX-99.1.

Chapter 11: Changing of the Guard

Interviews and Background Information

Bamforth, Mark
Carpenter, Robert
Christensen, Gustav
Collier, Duke
Coughlin, Bob
Csimma, Zoltan
Drake, Peter
Dunshire, Deborah
Dzau, Victor, *see* pages 138–139
Feldbaum, Carl
Ferrer, Carlos
Geraghty, James
Huyett, Bill
Johnson, Jack
Kaye, Ed
Lawton, Alison
Mack, Connie
Maderis, Gail
Maraganore, John
McDonough, Geoff
Meeker, David
Merrifield, Ann
Moscicki, Richard
Okita, Blair, *see* page 131
Ornskov, Flemming
Papadopoulos, Stelios
Piela, Bo
Poole, Sandra
Pops, Richard
Secor, Alicia
Sharp, Phil
Sibold, Bill
Skaletsky, Mark
Smith, Sandy
Tierney, Tomye, *see* pages 136–137
Valerio, Dinko
Viehbacher, Chris, *see* page 134
Walts, Alan
Webb, Lance
Wirth, Peter, *see* page 139
Wood, Joan
Zerhouni, Elias

References and Additional Readings

Anderson, Howard. 2010. Carl Icahn's battle to take down Genzyme, op-ed. *The Boston Globe*, April 13.
Everett W, Adams M. 2010. Keeping the life in life sciences, op-ed. *The Boston Globe*, April 13.
George W. 2010. Another view: can biotech survive Icahn? *New York Times* blog, June 3.

Chapter 12: The Diaspora

Interviews and Background Information

Albers, Jeff, *see* page 146
Been, Kees
Boger, Joshua
Butler, John, *see* pages 146–147
Carpenter, Robert
Christensen, Gustav, *see* page 141
Collier, Duke
Coughlin, Bob
Csimma, Zoltan
de los Pinos, Elisabet
de Boer, Daniel
Doliveux, Roch

Drake, Peter, *see* page 144
Dunshire, Deborah
Enyedy, Mark
Feldbaum, Carl
Geraghty, James
Greenwood, Jim
Huyett, Bill
Incerti, Carlo
Kaye, Ed
Lawton, Alison, *see* pages 145, 148–149
Maderis, Gail, *see* pages 142-143
Maraganore, John
McDonough, Geoff
Meeker, David, *see* pages 141, 147, 150
Merrifield, Ann
Moscicki, Richard
Ornskov, Flemming
Papadopoulos, Stelios
Phelps, Greg, *see* page 147
Piela, Bo
Poole, Sandra
Secor, Alicia
Sharp, Phil
Sibold, Bill
Slavin, Peter
Smith, Sandy
Soteropoulos, Paula, *see* pages 149–150
Valerino, Dinko
van Heek, Jan
Viehbacher, Chris
Vivaldi, Rogerio
Walts, Alan
Wirth, Peter
Wood, Joan
Xue, James

References and Additional Readings

Tirrell M. 2011. Genzyme chief's Sanofi deal payout as much as $221.2 million. https://www.bloomberg.com/news/articles/2011-03-098/genzyme-chief-s-sanofi-deal-payout-as-much-as-221-2-million

Weisman R. 2015. How Genzyme became a source of biotech executives. *The Boston Globe*, July 12.

General Background Source

Blanchard K, Diaz-Ortiz C. 2017. *One minute mentoring: how to find and work with a mentor—and why you'll benefit from being one.* HarperCollins, New York.

Sonenfeld J. 1988. *The hero's farewell: what happens when CEOs retire.* Oxford University Press, New York.

Chapter 13: A Citizen in Full

Interviews and Background Information

Bancel, Stéphane
Boger, Joshua
Boice, Nicole
Carpenter, Robert
Coughlin, Bob
Daley, George, *see* page 158
Dzau, Victor
Everett, Wendy
Geraghty, James
Greenwood, Jim
Haber, Daniel, *see* page 159
Hockfield, Susan, *see* pages 160–161
Howe, John
Minogue, Michael
Nissinen, Mikko, *see* pages 161–162
Patrick, Deval, *see* pages 163–164
Pyott, David
Reede, Joan, *see* page 157
Reif, Rafael, *see* page 160
Rosengren, Eric, *see* pages 162–163

Saltonstall, Peter
Sharp, Phil
Slavin, Peter, *see* page 158
Termeer, Belinda
Thier, Samuel, *see* page 159
Wirth, Peter

References and Additional Readings

U.S. Senate Congressional Record. 1999. Henri Termeer presented with the International Institute of Boston's Golden Door Award, p 29999. U.S. Government Publishing Office, Washington, DC.

Chapter 14: The Oracle of Marblehead

Interviews and Background Information

Bancel, Stéphane
Been, Kees
Boger, Joshua
Boice, Nicole
Carpenter, Robert
Collier, Duke
Csimma, Zoltan, *see* page 168
de los Pinos, Elisabet
de Boer, Daniel
Doliveux, Roch
Dunsire, Deborah
Enyedy, Mark
Everett, Wendy
Greenwood, Jim
Hillback, Elliott, *see* pages 165–166
Hockfield, Susan
Howe, John
Huyett, Bill
Kaye, Ed, *see* page 167
Maraganore, John, *see* page 171
McDonough, Geoff
Meeker, David, *see* page 169
Minogue, Michael
Ornskov, Flemming
Phelps, Greg
Piela, Bo
Pyott, David
Secor, Alicia
Sharp, Phil, *see* page 166
Sibold, Bill
Slavin, Peter
Smith, Sandy
Soteropoulos, Paula
Termeer, Belinda, *see* page 165–167
Valerio, Dinko
van Heek, Jan
Vivaldi, Rogerio
Walts, Alan
Wirth, Peter
Wood, Joan
Xue, James

Epilogue: Forgotten No More

Interviews and Background Information

Geraghty, James
Groft, Steve

References and Additional Readings

National Organization for Rare Disorders (NORD) Report. 2018. Orphan drugs in the United States: Growth trends in rare disease treatments. IQVIA Institute for Human Data Science.

https://www.iqvia.com/institute/reports/orphan-drugs-in-the-united-states-growth-trends-in-rare-disease-treatments
U.S. Government Accountability Office (GAO) Report to Congressional Requesters. 2018. Orphan drugs: FDA could improve designation review consistency; rare disease drug development challenges continue. https://www.gao.gov/products/GAO-19-83
Orphan Product Designations and Approval. https://www.ngocommitteerarediseases.org/

General Background Sources

Sasinowski FJ, Panico EB, Valentine JE. 2015. Quantum of effectiveness evidence in FDA's approval of orphan drugs: update, July 2010 to June 2014. *Ther Innov Reg Sci* **49:** 680–697. doi:10.1177/0092861511435906

General Background Sources for the Book

Deutsch CH. 1988. Staying alive in biotech: a struggling young company learns that bureaucracy is not all bad. *The New York Times*, Sunday, November 6.
Ghoshal S, Bartlett CA. 1988. *The individualized corporation: a fundamentally new approach to management. Great companies are defined by purpose, process, and people.* William Heinemann, Random House, London.
James FE. 1988. SmithKline to buy 9.6% Invitron stake in pact for manufacturing AIDS drug. *The Wall Street Journal*, Tuesday, September 6.
Steyer R. 1988. Protein PRETZELS: cell "factories" use a genetic twist to give drugs more potency. *Science Notes*, Tuesday, April 12.
Cogner JA. 1989. *The charismatic leader: behind the mystique of exceptional leadership.* Jossey-Bass, San Francisco.
Hawkins JD. 1991. *Gene structure and expression*, 2nd ed. Press Syndicate of the University of Cambridge, Cambridge.
Berg P, Singer M. 1992. *Dealing with genes: the language of heredity.* University Science Books, Mill Valley, CA.
Werth B. 1994. *The billion-dollar molecule: the quest for the perfect drug.* Simon & Schuster, New York.
Kotter JP. 1996. *Leading change.* Harvard Business School Press, Boston.
Bauman RP, Jackson P, Lawrence JT. 1997. *From promise to performance: a journey of transformation at SmithKline Beecham.* Harvard Business School, Boston.
Anand G. 2010. *The cure: how a father raised $100 million and bucked the medical establishment in a quest to save his children.* HarperCollins, New York.
Jones M, Dick B, Nelson H. 2012. *Honoring 25 years of biotech leadership: the Biotech Hall of Fame awards.* Life Sciences Foundation, San Francisco.
Bartlett CA, Khanna T, Choudhury P. 2012. *Genzyme's CSR dilemma: how to play its HAND.* Harvard Business School, Boston.
Termeer HA. 2014. A biotechnology entrepreneur's story: advice to future entrepreneurs. In *Biotechnology entrepreneurship*, Chap. 2, pp. 15–20. Elsevier, New York.
Brooks D. 2015. *The road to character.* Random House, New York.
Schramm CJ. 2018. *Burn the business plan: what great entrepreneurs really do.* Simon & Schuster, New York.

Genzyme. 2001. *A different vision: the making of Genzyme.* 20th anniversary book. Genzyme, Boston.

Conversations with Henri Termeer, oral history conducted by Ted Everson, Jennifer Dionisio, Pei Koay, and Arnold Thackray, May 23, December 7, 2006; August 2, 2007; December 18, 2008; and September 30, 2011; edited by Gavin Rynne and Mark Jones. © 2012 The Life Sciences Foundation, San Francisco.

Master List—Interviews

Albers, Jeff	Interviewed July 26, 2018
Aldrich, Richard	Interviewed December 11, 2018; December 12, 2018
Aliski, William	Interviewed January 7, 2019
Arnstein, Caren	Interviewed October 31, 2018
Bamforth, Mark	Interviewed December 4 2018
Bancel, Stéphane	Interviewed February 27, 2018
Barrell, Alan	Interviewed September 5, 2018
Barton, Norman	Interviewed October 3, 2018
Been, Kees	Interviewed August 1, 2018
Berger, Jacques	Interviewed May 31, 2018
Blair, Henry	Interviewed February 8, 2018
Boger, Joshua	Interviewed December 4, 2018
Boice, Nicole	Interviewed June 6 2018
Butler, John	Interviewed July 24, 2018
Buyers, Rhonda	Interviewed October 26, 2018
Carpenter, Robert	Interviewed February 7, 2018; October 11, 2018
Casamento, Chuck	Interviewed October 2, 2018
Castaldi, David	Interviewed October 11, 2018
Christensen, Gustav	Interviewed October 2, 2018
Collier, Duke	Interviewed March 1, 3028
Coolidge, Kathleen	Interviewed October 31, 2018
Cooney, Charles	Interviewed October 23, 2017
Coughlin, Bob	Interviewed August 2, 2018
Cox, Geoffrey	Interviewed February 6, 2018
Crowley, John	Interviewed October 29, 2018
Csimma, Zoltan	Interviewed February 22, 2018
Daley, George	Interviewed February 27, 2018
de Boer, Daniel	Interviewed May 28, 2018
de los Pinos, Elisabet	Interviewed August 30, 2018
Doliveux, Roch	Interviewed May 30, 2018
Drake, Peter	Interviewed January 22, 2018
Dunsire, Deborah	Interviewed July 25, 2018
Dzau, Victor	Interviewed March 13, 2018
Ely, Robin	Interviewed October 20, 2017

Enyedy, Mark	Interviewed March 21, 2018
Everett, Wendy	Interviewed December 17, 2018
Feldbaum, Carl	Interviewed May 9, 2018
Ferrer, Carlos	Interviewed December 10, 2018
Fleming, Dave	Interviewed April 18, 2018
Froehlich, Sara	Interviewed November 20, 2018
Gantz, Wilbur (Bill)	Interviewed October 3, 2018
Geraghty, James	Interviewed February 7, 2018; October 16, 2018
Greenwood, James (Jim)	Interviewed June 13, 2018
Groft, Steve	Not recorded
Haber, Daniel (Dan)	Interviewed July 26, 2018
Haffner, Marlene	Interviewed March 6, 2018
Heffernan, Jack	Interviewed February 22, 2018
Hillback, Elliott	Interviewed February 22, 2018; October 12, 2018
Hockfield, Susan	Interviewed June 6, 2018
Holcombe, Kay	Interviewed January 5, 2019
Howe, John	Interviewed March 14, 2018
Huyett, Bill	Interviewed March 1, 2018
Incerti, Carlo	Interviewed October 29, 2018
Johnson, Jack	Interviewed November 27 2018
Kaye, Edward (Ed)	Interviewed February 21, 2018
King, John	Interviewed November 29, 2018
Lawton, Alison	Interviewed January 23, 2018
Leonard, Dan	Interviewed November 29, 2018
Littlechild, John	Interviewed May 19, 2018
Lodish, Harvey	Interviewed October 27, 2017
Mack, Connie	Interviewed December 27, 2018
Maderis, Gail	Interviewed April 13, 2018
Maraganore, John	Interviewed August 1, 2018
McDonough, Geoffery (Geoff)	Interviewed August 1, 2018
McGrane, Mary	Interviewed November 6, 2018
McLachlan, David	Interviewed February 7, 2018
Meeker, David	Interviewed October 25, 2017
Meijer, Richard (Dick)	Interviewed December 6, 2018
Merrifield, Ann	Interviewed March 22, 2018
Meyers, Abbey	Interviewed January 29, 2018
Minogue, Michael	Interviewed December 15, 2018
Mistry, Pramod	Interviewed July 23, 2018
More, Eileen (Ginger)	Interviewed March 15, 2018
Moscicki, Richard	Interviewed January 31, 2018; January 10, 2019
Moufflet, Gerard	Interviewed October 4, 2018
Myers, Michael	Interviewed February 5, 2019
Nathan, Mary	Interviewed February 12, 2018

Nissinen, Mikko	Interviewed November 16, 2018
Okita, Blair	Interviewed September 14, 2018
Ornskov, Flemming	Interviewed October 24, 2018
Papadopoulos, Stelios	Interviewed March 26, 2018
Patrick, Deval	Interviewed October 31, 2018
Phelps, Gregory (Greg)	Interviewed February 6, 2018
Piela, Bo	Interviewed October 3, 2018
Poole, Sandra	Interviewed May 9, 2018; December 3, 2018
Pops, Richard	Interviewed July 25, 2018
Pyott, David	Interviewed February 6, 2018
Reede, Joan	Interviewed November 29, 2018
Reif, Rafael	Interviewed November 29, 2018
Ring, Jamie	Interviewed December 3, 2018
Rosenblatt, Michael (Mike)	Interviewed October 24, 2017
Rosengren, Eric	Interviewed December 4, 2018
Rubinstein, Linda	Interviewed March 1, 2018
Saltonstall, Peter	Interviewed November 1, 2018
Sasinowski, Frank	Interviewed November 8, 2018
Schmergel, Gabriel (Gabe)	Interviewed September 20, 2018
Schoneveld van der Linde, Maryze	Interviewed November 2, 2018
Secor, Alicia	Interviewed April 5, 2018
Shah, Uzma	Interviewed August 2, 2018
Sharp, Phillip (Phil)	Interviewed March 1, 2018
Sherblom, Jim	Interviewed April 4, 2018
Sibold, Bill	Interviewed March 21, 2018
Skaletsky, Mark	Interviewed October 4, 2018
Slavin, Peter	Interviewed November 1, 2018
Smith, Alan	Interviewed February 27, 2018
Smith, C. Ray	Interviewed April 2, 2018
Smith, Sanford (Sandy)	Interviewed March 29, 2018
Soteropoulos, Paula	Interviewed July 24, 2018
Taminiau, Ineke	Interviewed November 3, 2018
Taunton-Rigby, Alison	Interviewed March 20, 2018
Tayot, Jean-Louis	Interviewed May 31, 2018
Termeer, Belinda	Interviewed November 3, 2018; December 18, 2018
Termeer, Bert	Interviewed November 3, 2018
Termeer, Nicholas	Interviewed December 18 2018
Termeer, Paul	Interviewed November 3, 2018
Termeer, Roel	Interviewed November 3, 2018
Terry, Sharon	Interviewed December 1, 2018
Thier, Samuel	Interviewed December 17, 2018
Tierney, Tomye	Interviewed March 22, 2018

Valerio, Dinko	Interviewed May 28, 2018
van Heek, Jan	Interviewed April 2, 2018
Verduijn, Marlies	Interviewed November 3, 2018
Viehbacher, Christopher (Chris)	Interviewed August 2, 2018
Vivaldi, Rogerio	Interviewed February 28, 2018
Walts, Alan	Interviewed October 2, 2018
Webb, Lance	Interviewed October 12, 2018
Wirth, Peter	Interviewed March 21, 2018
Wood, Joan	Interviewed February 17, 2018
Xue, James	Interviewed May 9, 2018
Zerhouni, Elias	Interviewed July 18, 2018

Index

L

M